The IEE

Guidance Note 1

Selection & Erection

17th IEE Wiring Regulatior
BS 7671:2008 Requ ... stallations

Published by The Institution of Engineering and Technology, London, United Kingdom

The Institution of Engineering and Technology is registered as a Charity in England & Wales (no. 211014) and Scotland (no. SCO38698).

The Institution of Engineering and Technology is the new institution formed by the joining together of the IEE (The Institution of Electrical Engineers) and the IIE (The Institution of Incorporated Engineers). The new Institution is the inheritor of the IEE brand and all its products and services, such as this one, which we hope you will find useful. The IEE is a registered trademark of the Institution of Engineering and Technology.

First published 1992 (0 85296 535 4)
Reprinted (with amendments) 1993
Second edition (incorporating Amendment No. 1 to BS 7671:1992) 1996 (0 85296 864 7)
Third edition (incorporating Amendment No. 2 to BS 7671:1992) 1999 (0 85296 954 6)
Fourth edition (incorporating Amendment No. 1 to BS 7671:2001) 2001 (0 85296 989 9)
Reprinted (incorporating Amendment No. 2 to BS 7671:2001) 2004
Fifth edition (incorporating BS 7671:2008) 2009 (978-0-86341-855-6)

Copies of this publication may be obtained from:
The Institution of Engineering and Technology
PO Box 96
Stevenage
SG1 2SD, UK
Tel: +44 (0)1438 767328
Email: sales@theiet.org
www.theiet.org/publishing/books/wir-reg/

ISBN 978-0-86341-855-6

Typeset in the UK by The Institution of Engineering and Technology
Printed in the UK by Printwright Ltd, Ipswich

Contents

Cooperating organisations

The Institution of Engineering and Technology acknowledges the contribution made by the following organisations in the preparation of this Guidance Note.

Association of Manufacturers of Domestic Appliances
S.A. MacConnacher BSc CEng MIEE

British Cables Association
C.K. Reed IEng MIIE

British Electrotechnical & Allied Manufacturers Association Ltd
P. Galbraith IEng MIET MCMI
R.F.B. Lewington MIEE
M.H. Mullins BA CEng MIEE FIIE
P. Sayer IEng MIET GCGI

Electrical Contractors' Association
H.R. Lovegrove IEng FIET
Eur Ing L. Markwell MSc BSc CEng MIET MCIBSE LCGI

Electrical Safety Council
S. Hesketh IEng MIEE MAPEA CertEd(FE)
M.D. Link

GAMBICA Association Ltd
M. Hadley

Health and Safety Executive
K. Morton BSc CEng MIEE
N. Gove MEng CEng MIEE

Institution of Engineering and Technology
M. Coles (Editor) BEng(Hons) MIEE
G.D. Cronshaw IEng FIET
P.E. Donnachie BSc CEng FIET
J.F Elliott BSc(Hons) IEng MIEE
I.M. Reeve BTech CEng MIEE
Eur Ing G. Stokes BSc(Hons) CEng FIEE FCIBSE

Lighting Association
K.R. Kearney IEng MIET

SELECT (Electrical Contractors' Association of Scotland)
D. Millar IEng MIET MILE
M.M. Duncan IEng MIEE MILE

Acknowledgements

References to British Standards, CENELEC Harmonization Documents and International Electrotechnical Commission standards are made with the kind permission of BSI. Complete copies can be obtained by post from **BSI Customer Services**, 389 Chiswick High Road, London W4 4AL. Tel: 020 8996 9001, Fax: 020 8996 7001.

BSI operates an export advisory service – Technical Help to Exporters – which can advise on the requirements of foreign laws and standards. The BSI also maintains stocks of international and foreign standards, with many English translations.

Up-to-date information on BSI standards can be obtained from the BSI website: www.bsi-global.com

RCD terminology and information was provided by the Low Voltage Circuit Breaker Division of BEAMA and is taken from the BEAMA Code of Practice for the Selection and Application of Residual Current Devices.

Information on the forms of separation of LV switchgear were provided by the Low Voltage Distribution Switchboard Division of BEAMA.

The BEAMA booklet *Guide to forms of separation* can be obtained from **BEAMA**, Westminster Tower, 3 Albert Embankment, London SE1 7SL. Tel: 020 7793 3000.

Extracts from the LDSA document *Fire safety guide No. 1* on Section 20 buildings are reproduced with permission from The London District Surveyors Association, www.londonbuildingcontrol.org.uk

Guidance on cable separation in Appendix I was reproduced from the Electrical Contractors' Association document *Recommended cable separations to achieve electromagnetic compatibility (EMC) in buildings* with the permission of the ECA, ESCA House, 34 Palace Court, London W2 4HY.

The table on colour identification of buried services is reproduced with the permission of the **National Joint Utilities Group**, 30 Millbank, London SW1P 4RD.

The tables on cable references in Appendix F were reproduced with permission of **Anixter (UK) Ltd**, 1 York Road, Uxbridge, Middlesex UB8 1RN.

Copies of Health and Safety Executive documents and approved codes of practice (ACOP) can be obtained from **HSE Books**, PO Box 1999, Sudbury, Suffolk CO10 2WA. Tel: 01787 881165.

Preface

This Guidance Note is part of a series issued by the Institution of Engineering and Technology to explain and enlarge upon the requirements in BS 7671:2008, the 17th Edition of the IEE Wiring Regulations.

Note that this Guidance Note does not ensure compliance with BS 7671. It is intended to explain some of the requirements of BS 7671 but readers should always consult BS 7671 to satisfy themselves of compliance.

The scope generally follows that of BS 7671; the relevant Regulations and Appendices are noted in the margin. Some Guidance Notes also contain material not included in BS 7671:2008 but which was included in earlier editions of the Wiring Regulations. All of the Guidance Notes contain references to other relevant sources of information.

Electrical installations in the United Kingdom that comply with BS 7671 are likely to satisfy Statutory Regulations such as the Electricity at Work Regulations 1989 (EWR), however this cannot be guaranteed. It is stressed that it is essential to establish which Statutory and other Regulations apply and to install accordingly. For example, an installation in premises subject to licensing may have requirements different from, or additional to, BS 7671 and these will take precedence.

Introduction

This Guidance Note is concerned principally with Part 5 of BS 7671 – Selection and erection of equipment.

Neither BS 7671 nor the Guidance Notes are design guides. It is essential to prepare a full design and specification prior to commencement or alteration of an electrical installation. Compliance with the relevant standards should be required.

The design and specification should set out the requirements and provide sufficient information to enable competent persons to carry out the installation and to commission it. A good specification must include a description of how the system is to operate and all the design and operational parameters. It must provide for all the commissioning procedures that will be required and for the provision of adequate information to the user. This will be by means of an operational manual or schedule and 'as fitted' drawings if necessary. 514.9

It must be noted that it is a matter of contract as to which person or organisation is responsible for the production of the parts of the design, specification, installation and any operational information. The persons or organisations who may be concerned in the preparation of the works include:

The Designer
The Planning Supervisor
The Installer (or Contractor)
The Supplier of Electricity (Distributor)
The Installation Owner (Client) and/or User
The Architect
The Fire Prevention Officer
Any Regulatory Authority
Any Licensing Authority
The Health and Safety Executive

In producing the design and specification, advice should be sought from the installation owner and/or user as to the intended use. Often, as in a speculative building, the intended use is unknown. In such cases the specification and/or the operational manual must set out the basis of use for which the installation is suitable. 132.1

Precise details of each item of equipment specified should be obtained from the manufacturer and/or supplier and compliance with appropriate standards confirmed. 133.1.1 511

The operational manual must include a description of how the system as installed is to operate and all commissioning records. The manual should also include manufacturers' technical data for all items of electrical equipment, wiring, switchgear, accessories, etc. and any special instructions that may be needed. The Health and Safety at Work etc. Act 1974 (HSWA) Section 6 and the Construction (Design and Management) Regulations 2007 (CDM) are concerned with the provision of information. Guidance on the preparation of technical manuals is given in BS 4884 and BS 4940. The size and complexity of the installation will dictate the nature and extent of the manual.

General requirements 1

1.1 General

'Equipment' is short for 'electrical equipment' and is defined as: Part 2

> *Any item for such purposes as generation, conversion, transmission, distribution or utilization of electrical energy, such as machines, transformers, apparatus, measuring instruments, protective devices, wiring systems, accessories, appliances and luminaires.*

Installation designers may need information from the manufacturer as to the suitability of equipment for its intended use. It is not the intention of BS 7671:2008 to stifle innovation or new techniques but, generally, the standard only recognises and considers established materials and techniques. 120.2 133.1.1 133.1.2 120.4 511.1 511.2

1.2 Equipment

The designer is responsible for the safety of the design. Any intended departures from BS 7671, although the designer is confident regarding safety, must be recorded on the Electrical Installation Certificate. The resulting degree of safety of the installation must not be less than that obtained by compliance with the Regulations. 120.3

Chapter 13 of BS 7671, 'Fundamental Principles', outlines the basic requirements. Later chapters describe in more detail particular means of compliance with Chapter 13. Chapter 13 is normally referred to only where it is intended to adopt a method not recognised in later chapters. The terms are chosen to allow for interpretation to suit special cases. Chap 13

The phrase *so far as is reasonably practicable* is used in Parts 6 and 7 of the Regulations. It should be borne in mind that methods described in later chapters are considered reasonably practicable in most circumstances. Part 6 Part 7

Where methods are used which are not described in later chapters, then the onus is on the designer to confirm that the degree of safety is not less than that required by Chapter 13.

It is important to note that Approved Document P, which has been approved and issued by the responsible Secretary of State for the purpose of providing practical guidance with respect to Schedule 1 and Regulation 7 of the Building Regulations for England and Wales, states that the requirements of Part P (applicable to dwellings) will be met by adhering to Chapter 13, 'Fundamental Principles', of BS 7671.

1.3 The Electricity at Work Regulations 1989

The requirements of the Electricity at Work Regulations 1989 (EWR) are intended to provide for the safety of persons gaining access to or working with or near electrical equipment.

The Memorandum of Guidance on the Electricity at Work Regulations 1989 (Health and Safety Executive publication (HSR25) should be carefully studied and it should be borne in mind that the EWR apply to designers, installers and users of installations alike.

BS 7671 is intended for designers, installers and those verifying electrical installations. The Regulations, therefore, include requirements for the design of the installation and
132.12 the selection and erection of electrical equipment. The installation designer should
Chap 34 assess the expected maintenance of the installation and the initial design should make provision for maintenance to be carried out. The user has the responsibility for ensuring that equipment is properly operated and maintained when necessary.

Regulation 4(2) of the EWR requires that adequate maintenance is undertaken to prevent danger and HSR25 advises that regular inspection of electrical systems (supplemented by testing as necessary) is an essential part of any preventive maintenance programme. Regular operational functional testing of safety circuits (emergency switching/stopping, etc.) may be required, as, unlike functional circuits, they may be infrequently used. Comprehensive records of all inspections and tests should be made, retained and reviewed for any trends that may arise.

Guidance Note 3: *Inspection & Testing* gives guidance on initial and periodic inspection and testing of installations.

1.4 The Construction (Design and Management) Regulations 2007

The Construction (Design and Management) Regulations 2007 (CDM Regulations) place responsibilities on most installation owners and their professional design teams to ensure a continuous consideration of health and safety requirements during the design and construction of, and throughout the life of, an installation, including maintenance, repair and demolition.

Almost everyone involved in construction work will have a legal duty placed on them under the regulations. Those with legal duties are commonly known as 'dutyholders'.

Dutyholders under the CDM Regulations are:

- **Clients** – A 'client' is anyone having construction or building work carried out as part of their business. This could be an individual, partnership or company and includes property developers or management companies for domestic properties.
- **CDM coordinators** – A 'CDM coordinator' has to be appointed to advise the client on projects that involve more than 30 days or 500 person days of construction work. The CDM coordinator's role is to advise the client on health and safety issues during the design and planning phases of construction work.

- **Designers** – The term 'designer' has a broad meaning and relates to the function performed, rather than the profession or job title. Designers are those who, as part of their work, prepare design drawings, specifications, bills of quantities and the specification of articles and substances. This could include architects, engineers and quantity surveyors.
- **Principal contractors** – A 'principal contractor' has to be appointed for projects which last more than 30 days or involve 500 or more person days of construction work. The principal contractor's role is to plan, manage and coordinate health and safety while construction work is being undertaken. The principal contractor is usually the main or managing contractor for the work.
- **Contractors** – A 'contractor' is a business who is involved in construction, alteration, maintenance or demolition work. This could involve building, civil engineering, mechanical, electrical, demolition and maintenance companies, partnerships and the self-employed.
- **Workers** – A 'worker' is anyone who carries out work during the construction, alteration, maintenance or demolition of a building or structure. A worker could be, for example, an electrician as well as those supervising the work, such as chargehands and foremen.

1.5 Building Regulations

1.5.1 The Building Regulations 2000 (England and Wales)

A series of Approved Documents have been published for the purpose of providing practical guidance to the requirements of the *Building Regulations 2000 (SI 2000/2531) for England and Wales*. Approved Documents of particular interest to designers and installers of electrical installations are:

- A – Structure
- B – Fire safety, Volume 1: Dwellinghouses
- B – Fire safety, Volume 2: Buildings other than dwellinghouses
- C – Site preparation and resistance to contaminants and moisture
- E – Resistance to the passage of sound
- F – Ventilation
- L1A – Conservation of fuel and power (New dwellings)
- L1B – Conservation of fuel and power (Existing dwellings)
- L2A – Conservation of fuel and power (New buildings other than dwellings)
- L2B – Conservation of fuel and power (Existing buildings other than dwellings)
- M – Access to and use of buildings
- P – Electrical safety: Dwellings
- Regulation 7 – Materials and workmanship.

Approved Document B includes a requirement for the maintenance of a fire barrier.

Approved Document C includes a requirement for maintenance of airtightness; see also BS 9250:2007 *Code of Practice for the airtightness of ceilings in pitched roofs.*

Approved Document E includes a requirement for the maintenance of acoustic resistance.

Approved Documents L1A and L1B contain a requirement that reasonable provision shall be made for occupiers to obtain the benefits of efficient lighting.

Approved Document M includes requirements for the mounting heights of accessories and these are referred to in Appendix C of this Guidance Note (see C.7).

Internal lighting

One way of demonstrating compliance with Approved Document L1A would be to install fixed lighting that only takes lamps having a luminous efficacy greater than 40 lumens per circuit watt:

a one per 25 m^2 of dwelling floor area (excluding garages), or
b one per four fixed light fittings,

whichever is the greater.

Examples of lamps which achieve this efficacy are fluorescent lamps and compact fluorescent lamps.

External lighting

External lighting includes lighting in porches but does not include lighting in garages or carports. Reasonable provision shall be made to enable effective control such that either:

a the lamp capacity does not exceed 150 W per light fitting and the light automatically switches off when:
  i there is sufficient daylight, and
  ii it is not required at night, or
b the fittings have lampholders that can only be used with lamps having an efficacy greater than 40 lumens per watt.

▼ **Figure 1.1**
External luminaire incorporating photo-cell and PIR control. Lamp rating not exceeding 150 W.

Approved Documents L2A and L2B contain requirements for general lighting efficacy in offices, industrial and storage buildings. Lighting systems in such buildings shall be provided with reasonably efficient lamp/luminaire combinations. A way of complying with the requirements would be to provide lighting with an initial efficacy, averaged over the whole building, of not less than 40 luminaire-lumens per circuit watt. Approved Document L also includes requirements for display lighting.

For further information, the Approved Documents should be consulted.

1.5.2 The Building (Scotland) Regulations 2004

The Building Regulations 2000 are not applicable in Scotland, where the Building (Scotland) Regulations 2004 apply.

The detailed requirements are given in the Technical Standards for compliance with the Building (Scotland) Regulations.

Guidance on how to achieve compliance with these Standards is given in two Scottish Building Standards Technical Handbooks – Domestic and Non-domestic.

These handbooks contain recommendations for electrical installations including the following:

- Compliance with BS 7671
- Minimum number of socket-outlets in dwellings
- Minimum number of lighting points in dwellings
- Minimum illumination levels in common areas of domestic buildings, for example, blocks of flats
- Minimum mounting heights of switches and socket-outlets, etc.
- Separate switching for concealed socket-outlets, for example, behind white goods in kitchens
- Conservation of fuel and power in both domestic and non-domestic buildings.

With regard to electrical installations in Scotland, the requirements of the above Building Regulations are deemed to be satisfied by complying with BS 7671.

The handbooks may be obtained from the Scottish Building Standards Agency.

Web: www.sbsa.gov.uk
Telephone orders: 01506 600400
Email orders: info@sbsa.gsi.gov.uk

1.6 Competence

BS 7671 requires that an assessment be made of the external influences to which the installation is exposed. Amongst these, as indicated in Appendix 5 of BS 7671, is Category BA – classification of the capability of persons. This recognises different levels, from an ordinary person to a skilled person. It must be remembered that a skilled person will not be skilled in all facets of electrical work and consequently will not be competent in all works. 301.1 Appx 5

BS 7671 defines a competent person as: Part 2

> *A person who possesses sufficient technical knowledge, relevant practical skills and experience for the nature of the electrical work undertaken and is able at all times to prevent danger and, where appropriate, injury to him/herself and others.*

621.2 Persons performing electrical work must have sufficient technical knowledge or experience or be adequately supervised such that danger or injury is prevented. The object is that persons dealing with electrical equipment are not placed at risk due to a lack of skills on their own part, or on the part of others. Technical knowledge or experience may include basic knowledge of electricity, experience of electrical work, understanding of the system to be worked on, understanding of the hazards and the necessary precautions and the ability to recognise at all times whether it is safe for work to continue. Competence also includes a knowledge of safety in addition to safe
134.1.1 working practice and practical skills where needed.

Part 2 A person possessing technical knowledge or sufficient experience to enable him/her to avoid danger is recognised by BS 7671 as a 'skilled person' and a person adequately advised or supervised by one or more skilled persons to enable him/her to avoid the dangers that electricity may create is an 'instructed person'.

EWR Regulation 16 To require, or allow, a person who is not competent to undertake electrical work may be a breach of the statutory Health and Safety legislation, including the Electricity at Work Regulations 1989.

Selection and erection of equipment

2

2.1 Selection and erection of equipment

Chapter 51 gives the basic common rules to which every installation shall conform. 510.1

2.2 Compliance with standards

Part 3
Section 512

BS 7671 recognises equipment which complies with a British Standard or Harmonized Standard appropriate to the intended use of the equipment without further qualification. 511.1

A Harmonized Standard is defined in Part 2 as: Part 2

A standard which has been drawn up by common agreement between national standards bodies notified to the European Commission by all member states and published under national procedures.

There is a statutory definition of Harmonised Standard in Health and Safety Statutory Instrument 1992 No. 3073, the Supply of Machinery (Safety) Regulations 1992. This states:

'Harmonised standard' means a technical specification adopted by the European Committee for Standardisation (CEN) or the European Committee for Electrotechnical Standardisation (CENELEC) or both, upon a mandate from the Commission in accordance with Council Directive 83/189/EEC of 28 March 1983 laying down a procedure for the provision of information in the field of technical standards and regulations, and of which the reference number is published in the Official Journal of the European Communities.

2.2.1 European norms (ENs)

European Norm (EN) standards are standards that are required to be adopted by all CENELEC members and to be published with identical text by all members, without any additions, deletions or further technical amendments. Such EN standards then supersede the relevant national standards which are withdrawn to an agreed timescale. They are published as BS ENs in the UK. Appx 1

2.2.2 Harmonization documents (HDs)

Harmonization Documents (HDs) forming the CENELEC 384 and, more recently, 60364 series are documents based on the IEC 60364 suite of standards.

Harmonized Documents are standards that have been agreed by all CENELEC member countries and National Committees are obliged to implement those standards.

A National Standard based on an HD may have further technical additions, made by the national standards committee, but requirements are not to be deleted. Note that National Standards are not to be in conflict with the HD.

BS 7671 is based on a number of CENELEC HDs within the 384/60364 series (see the Preface to BS 7671) with extra specific technical material added.

Equipment is sometimes satisfactory only when used in a particular way or with other matching equipment. Certain equipment complying with a foreign standard may be safe when used, for example, with a foreign wiring system but may not be safe when used in conjunction with traditional UK practice.

511.1 511.2 Where equipment complying with a foreign standard which in turn is based on an IEC standard is specified, the designer or specifier must verify that the equipment is at least as safe as similar equipment complying with the relevant British or Harmonized Standard. It is the designer's responsibility to ensure that equipment they specify, which is not to a British or Harmonized Standard, provides the required performance and degree of protection. Note that BS 7671 does not insist on approval or certification of the equipment to the relevant Standard but this may be required in some cases by legislation or by the client.

2.3 Operational conditions and external influences

133 Equipment of all types must be suitable for its situation and use.

Part 3 Section 512 An assessment by the designer of the characteristics and conditions of the installation will be necessary, including the requirements of Part 3 of BS 7671:2008.

331.1 The installation must be designed to be suitable for all the relevant conditions and external influences foreseen, including electricity supply effects or effects on the supply.

Chap 32 Chap 52 512.2 Appx 5 Appendix 5 of BS 7671 details the system of classification of external influences developed in IEC 60364-3 and the classification is indicated in parts of Chapter 52. While not in general use in the UK, the classification can serve as a helpful reminder of conditions to be considered.

331.1 Regulation 331.1 lists some examples of characteristics of equipment which may adversely affect the equipment to be installed, the electricity supply or other services.

This list is not exhaustive but shows some details of what the designer should consider. Equipment must operate safely; further, it should be efficient and correctly selected.

510.1 All equipment must be selected to accommodate the worst foreseeable conditions of service that can be encountered even if such conditions happen rarely.

The assessment of general characteristics requires a thorough consideration of the purpose for which the installation is intended to be used. Further, the building, installation structure and the electricity supply characteristics, including voltage, current and frequency, must be noted during the first part of the design. 301.1 512.1.1 512.1.2 512.1.3

If, for example, there is doubt that a switch or circuit-breaker can be used with inductive or capacitive circuits, e.g. motors, transformers or fluorescent lighting, advice should be obtained from the manufacturer. 512.1.4

2.3.1 Motors

Electric motors may have similar power ratings but differing applications. Motors for lifts, industrial plant and machinery, propulsion or ventilation will have differing duty cycles and will usually be the subject of a Standard. 512.1.4

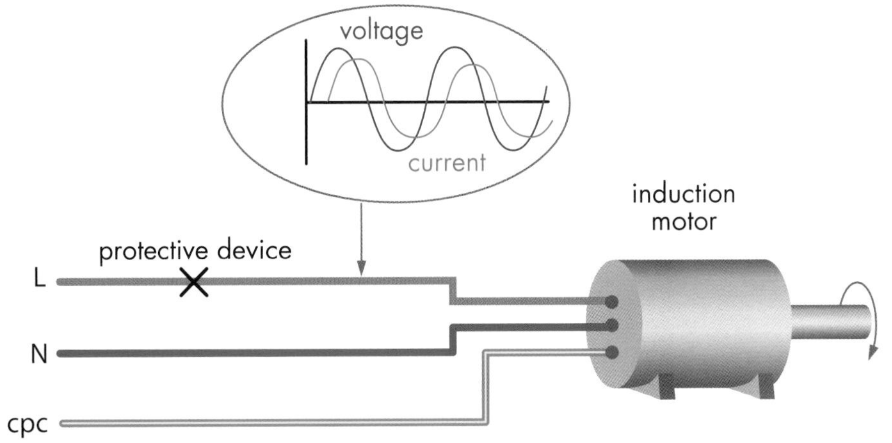

▼ Figure 2.1
Induction motor

The fixed wiring must be able to match the duty cycle of the connected load. Infrequently used motor-driven equipment with brakes will have different demands on the fixed wiring when compared with, for example, a hydro-extractor that uses 'plugging' as a method of braking the equipment frequently. The demands on the fixed wiring should be established from the manufacturer's installation instructions. These details will form part of the installation manual. (See Guidance Note 6 for detailed information on cable selection for motor circuits.) 552.1.1

2.3.2 Electromagnetic compatibility (EMC)

All equipment must be selected and erected to permit safe working, prevent harmful effects to other equipment and not impair the supply arrangements. This includes the consideration of electromagnetic compatibility (EMC) effects, as well as more straightforward considerations such as loading, current and voltage rating, circuit arrangements, etc. Regulation 515.3 specifically requires consideration of EMC when selecting and installing equipment. Both electromagnetic immunity (EMI), the protection of equipment from other electromagnetic influences (Regulation 515.3.1), and electromagnetic emission, the production and emission of electromagnetic interference by the proposed equipment (Regulation 515.3.2), shall be considered. This is a specialist application and advice from the equipment manufacturer should generally be taken. (See Appendix I for further information on cable separation.) 515.3

2.3.3 Harmonics

Harmonics are an aspect of EMC. Harmonic voltages and currents can cause interference with the normal operation of equipment and overload cables in certain cases (see Section 2.5 and Chapter 6 also).

2.4 Identification

2.4.1 Cables and conductors

Table 51 The means of identification for cables and conductors for fixed wiring are given in Table 51 of BS 7671 – reproduced here as Table 2.1 for convenience. Identification may be
514.3.1 by colour and/or alphanumeric marking. It is better to select cables having the correct colour identification wherever possible.

514.3.2 Cores of cables should be correctly identified at their terminations, and preferably throughout their length.

514.4.2 The colour combination green-and-yellow is reserved exclusively for identification of protective conductors and must not be used for any other purpose. Bare conductors and busbars which are to be used as protective conductors may be identified using green-and-yellow sleeving or tape, as appropriate.

521.9.3 Flexible cables and flexible cords may be used for fixed wiring, subject to the relevant provisions of the Regulations being met. These provisions are generally considered to relate to physical installation criteria.

▼ **Table 2.1**
Identification of cables and conductors (from Table 51 of BS 7671)

Function	Alphanumeric	Colour
Protective conductors		Green-and-yellow
Functional earthing conductor		Cream
a.c. power circuit[1]		
Line of single-phase circuit	L	Brown
Neutral of single- or three-phase circuit	N	Blue
Line 1 of three-phase a.c. circuit	L1	Brown[5]
Line 2 of three-phase a.c. circuit	L2	Black[5]
Line 3 of three-phase a.c. circuit	L3	Grey[5]
Two-wire unearthed d.c. power circuit		
Positive of two-wire circuit	L+	Brown
Negative of two-wire circuit	L-	Grey
Two-wire earthed d.c. power circuit		
Positive (of negative earthed) circuit	L+	Brown
Negative (of negative earthed) circuit[2]	M	Blue
Positive (of positive earthed) circuit[2]	M	Blue
Negative (of positive earthed) circuit	L-	Grey
Three-wire d.c. power circuit		
Outer positive of two-wire circuit derived from three-wire system	L+	Brown
Outer negative of two-wire circuit derived from three-wire system	L-	Grey
Positive of three-wire circuit	L+	Brown
Mid-wire of three-wire circuit[2,3]	M	Blue
Negative of three-wire circuit	L-	Grey
Control circuits, ELV and other applications		
Line conductor	L	Brown, Black, Red, Orange, Yellow, Violet, Grey, White, Pink or Turquoise
Neutral or mid-wire[4]	N or M	Blue

Notes to Table 2.1:

1 Power circuits include lighting circuits.
2 M identifies either the mid-wire of a three-wire d.c. circuit, or the earthed conductor of a two-wire earthed d.c. circuit.
3 Only the middle wire of three-wire circuits may be earthed.
4 An earthed PELV conductor is blue.
5 Appendix 7 of BS 7671, item 5 states that in a two- or three-phase power circuit the line conductors may all be one of the permitted colours, either identified L1, L2, L3 or marked brown, black, grey at their terminations to show the line designation.

2.4.2 Circuits

A diagram, chart or table giving details of the circuits is required. The details must be used by the person verifying compliance with BS 7671; for example, details of the selection and characteristics of the protective devices are needed for verifying protection against overcurrent and electric shock. A durable copy of the details should be fixed in or adjacent to each distribution board. The same details will be needed to assist periodic inspection and testing at a later date. 514.9.1 514.8.1 621.1

For simple installations, the Electrical Installation Certificate, together with the Schedule of Test Results and the Schedule of Inspections, will meet the requirements provided each circuit is also identified at the distribution board. 631.1 632.1 632.2

2.4.3 Switchgear

Labelling of switchgear is very important, particularly where the route of the final circuit cables is not obvious. If there is a possibility of confusion, some reliable means of identification must be clearly visible. It is necessary for the protective devices to be marked so that they can be identified easily by the user (see BS EN 60439). 514.1.1 514.8.1

Complex installations demand greater detail. Details of protective measures and cables should be provided as part of the 'as installed' information. When the occupancy of the premises changes, the new occupier should have sufficient information to correctly operate the electrical installation. Diagrams, charts, tables and schedules should be kept up to date. Such items are essential aids in the maintenance and periodic inspection and testing of an installation. 514.1.1 621.1

2.4.4 Warning notices

The warning notices called for in Section 514 of BS 7671 are intended to warn persons about the risk of working on or near live parts.

Warning notice – Voltage

The warning notices regarding voltage concern three different situations: 514.10.1

1 **Inside an item of equipment or enclosure**
Where the nominal voltage (U_0) within an item of equipment or enclosure exceeds 230 volts (e.g. a non-standard voltage such as 650 V) and where the presence of such a voltage would not normally be expected, before access is gained to a live part, there must be a clearly visible warning of the maximum voltage present.

2 **Between simultaneously accessible items of equipment or enclosures**
Where simultaneously accessible terminals or other fixed live parts have a nominal voltage (U_0) between them which exceeds 230 V (such as where the voltages are from different systems), then a warning notice must be fixed in such a position that anyone, before gaining access to such live parts, is warned of the maximum voltage which exists between those parts.

3 Where different nominal voltages exist

The means of access to all live parts of switchgear and other fixed live parts where different nominal voltages exist (e.g. different nominal voltages such as 110 V and 230 V present within switchgear) shall be marked to indicate the voltages present.

Note: Three-phase 400 V (U) systems have a nominal line to Earth voltage (U_0) of 230 V and do not normally require a warning notice.

2.4.5 Other notices

Other notices required by Section 514 are:

514.11.1 ▶ in each position where there are live parts which are not capable of being isolated by a single device
514.12.1 ▶ regarding periodic inspection and testing
514.13.1 ▶ regarding earthing and bonding connections
514.14.1 ▶ where wiring to two different colour systems is present
514.15.1 ▶ where the installation has dual supplies.

418.2 418.3 514.13.2 Where protection by earth-free local equipotential bonding or by electrical separation is used, the protective bonding conductors shall not be connected to earth. A warning notice to this effect, as specified in Regulation 514.13.2, must be fitted.

2.4.6 Safety signs

Item 9 of Appendix 2 of BS 7671 refers to the provision of safety signs as required by the Health and Safety (Safety Signs and Signals) Regulations 1996. Risks should be minimized at the design stage of an installation and warning signs only used where other methods of avoiding danger are not practical to implement.

2.4.7 Notices and identification labels

Table 2.2 lists the requirements for notices and identification labels given in BS 7671. Notices and labels should be of a size and type suitable for the location and installed such that they will not be painted over or easily removed or defaced. Labels, etc., should be permanently fixed, e.g. by suitable screws, rivets or resin glues, taking care not to damage equipment, invalidate IP ratings or block vents. Self-adhesive labels are not advisable and should only be used where heat or damp is not expected, as they cannot be considered permanent.

2.4.8 Conduit

514.2.1 Where an electrical service is required to be distinguished from a pipeline or another service, orange is to be used as the basic identification colour.

Notice, label or identification	Regulation
Socket-outlet labelled for connection of a particular item of equipment where additional protection by means of an RCD in accordance with Regulation 415.1 may be omitted	411.3.3(b)
Equipment retaining electrical charge after disconnection from the supply whilst behind a barrier or in an enclosure, e.g. capacitor	416.2.5
Protection by earth-free local equipotential bonding	418.2.5 514.13.2
Protection by electrical separation	418.3 514.13.2
Purpose of switchgear and controlgear (if not obvious)	514.1.1
Conduit, identification, where required	514.2.1
Bare protective conductors	514.4.2
Diagrams, charts, etc., for installation information	514.9.1
Marking nominal voltage present on means of access where exceeding 230 V	514.10.1
Different nominal voltages present	514.10.1
Live parts not capable of being isolated by a single device	514.11.1 537.2.1.3
Periodic inspection and testing	514.12.1
Quarterly test of a residual current device	514.12.2
Earthing and bonding connections	514.13.1 542.3.2
Earth-free equipotential bonding	514.13.2
Installation has wiring colours to two versions of BS 7671	514.14.1
High protective conductor current	514.16 543.7.1.5
Dual supply or supplies from different sources or circuits	514.15.1 537.1.6 537.2.1.3
Cables buried in the ground	522.8.10
Switching device for mechanical maintenance (indication of operation)	537.3.2.2
Firefighter's switch	537.6.4
Maximum current to be supplied from a temporary supply unit	559.10.7.2
Exemption from notices for periodic inspection and testing and testing of RCDs for outdoor lighting and highway installations	559.10.7.1
Drawings of electrical safety installations to be displayed at the origin of the installation	560.7.10
Notice warning users of equipment to use it only when swimming pool not occupied	702.410.3.4.1
For automatic life support for high density livestock rearing, a testing notice should be placed adjacent to the standby electrical source	705.560.6 (iii) a)
Socket-outlets at marinas – berthing instructions for connection to shore supply	709.553.1.10
PV array and generator junction boxes labelled to warn that the equipment may still be live after isolation	712.537.2.2.5.1
Permitted characteristics and type of supply for mobile and transportable units	717.514
Caravan instructions for supply and periodic inspection	721.514.1
Notice inside caravan for connection to supply	721.537.2.1.1.1
Plan for heating system to be fixed at distribution board	753.514

▼ **Table 2.2**
Notices and identification labels required by BS 7671

2.5 Mutual detrimental influence

132.11 Regulation 515.1 requires that there be no harmful effect between electrical and other
515.1 installations. The best approach, where practicable, is to arrange that the installations
515.2 are kept separated. Damage can be caused by such influences as thermal effects,
528.3 electrolysis or corrosion. The thermal effects of other installations, such as hot water systems, must be considered. Equipment should be either designed to operate properly at elevated temperatures, reduced temperatures, etc. or be protected from such effects. Electrolysis may result from leakage currents or from contact between dissimilar metals in damp conditions. Corrosion may result, for example, in rusting of unprotected steelwork; action needs to be taken to obviate these risks.

The Electromagnetic Compatibility (EMC) Regulations 2005 place a statutory requirement on designers and constructors to design/construct electrical equipment and systems so that they do not cause excessive electromagnetic interference
515.3 (emissions) and are not unduly affected by electromagnetic interference from other electrical equipment or systems. Harmonic production and interference, electrostatic discharges, mains-borne signals, etc. are all types of EM interference to be considered but the subject is too wide for this Guidance Note and requires specialist advice (see Chapter 7 and Appendix I for some further information).

The protection provided should form a prominent feature of the operational manual (see the Introduction to this Guidance Note).

2.6 Compatibility

133.4 All equipment should be selected and erected so that it will not be susceptible to
331.1 interference from, or cause harmful effects to, other equipment when used for the
512.1.5 purpose for which it is intended nor impair the supply during normal service, including switching operations (see sections 2.3 and 2.5).

331.1 Regulation 331.1 is of particular importance when considering supplies to information technology equipment; this includes computers, electronic office equipment, data transmission equipment and point-of-sale financial transaction terminals. Switch-mode power supply units, used to power this type of equipment, are particularly well known for causing harmonic generation problems in addition to high protective conductor currents.

Whenever electrical equipment is switched on or off, particularly where inductive loads are involved, high frequency voltage transients occur. This so-called mains-borne noise may cause malfunction in information technology equipment. The capacitance of the circuit cables and the filters incorporated in most information technology equipment attenuate this mains-borne noise. It is good practice to ensure that sources likely to give rise to significant noise, e.g. motors and thermostatically controlled equipment, are kept apart from sensitive equipment. Susceptible electronic equipment should be fed by separate circuits from the incoming supply point of the building. Additional filters (sometimes called 'power conditioners') may be used to reduce this transient noise on existing circuits and surge protective devices can be installed to divert or absorb transients (see section 3.9).

Large switch-on (inrush) current surges occur when transformers, motors, mains rectifier circuits, etc. are energized and can cause excessive short-time voltage drop in the circuit conductors (dips). Inrush currents affect other circuits also and may require larger cable sizes or the equipment to be put on separate circuits. It should be borne in mind that information technology equipment itself can cause the same problem if switched on in large groups.

Where maintenance of the supply voltage to especially sensitive equipment, such as information technology equipment, is considered of importance, the user may need a device such as a motor-generator or uninterruptible power supply (UPS).

2.7 Functional earthing

When planning installations consisting of information technology equipment, a dedicated conductor provided for functional earthing could be considered. Information on functional earthing is contained in Guidance Notes 5 and 8.

The protective conductors of the installation, which are connected to the main earthing terminal, are subject to voltages relative to the general mass of Earth, of which two types of voltage appearing on the earthing system can be considered.

The first is *transients.* These are short-duration spikes on the voltage waveform which can be caused by load switching of high voltage (HV) equipment or lightning strikes. Transients may be generated by the charging of an equipment frame via the stray capacitance from the mains circuit or mains-borne transients may be coupled into the earthing conductor or frame from mains conductors.

The second type of voltage is created by switch-mode power supplies, which are designed to filter current to earth; Ohm's law dictates the size of the induced voltage.

These voltages are termed *earth noise*, as any unwanted signal, or interfering voltage, is deemed to be 'noise'. Noise in the earthing system can cause disruption to sensitive microprocessors of computing equipment, as the earth-reference point is no longer 'zero'.

Manufacturers or designers of large computer systems usually make specific recommendations for the provision of a 'clean' mains supply or a 'clean' earth. The equipment manufacturer's guidance must be taken for such installations.

A dedicated earthing conductor may be used for a computer system, provided that:

1 all accessible exposed-conductive-parts of the computer system are earthed, the computer system being treated as an 'installation' where applicable 542.4.1

2 the main earthing terminal or bar of the computer system ('installation') is connected directly to the building main earthing terminal by a protective conductor. 543.1

2.8 Low voltage (LV) switchgear and controlgear assemblies

BS EN 60439-1:1999 gives guidance on the forms of separation applicable to factory-built switchgear and controlgear assemblies (switchboards, motor control centres, distribution boards, busbar trunking systems, etc.), known as type-tested and partially type-tested assemblies. These forms of separation provide protection against contact with live parts belonging to adjacent devices and protection from the probability of initiating arcing faults and the passage of foreign bodies between units of the assembly. The Standard also gives guidance on other requirements for protection against electric shock.

Definitions given in BS EN 60439-1 are:

Type-tested low voltage switchgear and controlgear assembly (TTA). A low voltage switchgear and controlgear assembly conforming to an established type or system without deviations likely to significantly influence the performance from the typical ASSEMBLY verified to be in accordance with this standard.

For various reasons, certain steps of assembly may be made in a place outside the factory of the manufacturer of the type-tested ASSEMBLY. Such an ASSEMBLY is considered as a type-tested ASSEMBLY provided the site assembly is carried out in accordance with the manufacturer's instructions, which should be retained for future information. The instructions will include any requirements for submission of the ASSEMBLY to further applicable routine test.

Partially type-tested low voltage switchgear and controlgear assembly (PTTA). A low voltage switchgear and controlgear assembly, containing both type-tested and non-type-tested arrangements provided that the latter are derived (e.g. by calculation) from type-tested arrangements which have complied with the relevant tests.

It should be noted that the forms of separation have no bearing on the overall ingress protection (IP rating or classification) for the external frame and enclosure of the switchgear assembly, etc. This should be specified to the required IP rating (see Appendix B).

BS EN 60439-1:1999 is identical to IEC 60439-1:1999 with the exception of the additional information given in National Annex NA. Designers and specifiers working on international projects should clarify specific requirements with the manufacturer.

2.8.1 Forms of separation

Four forms of separation are indicated in the main text of BS EN 60439-1:1999 but there is no specific detail given on how these forms are to be achieved. It is stated in the Standard that the form of separation should be agreed between manufacturer and designer/user. It shall be remembered that higher forms of separation specified will increase costs but will give better operational flexibility regarding safe working when connecting in additional circuits or carrying out maintenance. This 'trade-off' must be carefully assessed.

The four forms given have basic definitions and applications but Forms 2 to 4 can be further subdivided into more specific 'Types' (applications) by discussion and agreement with manufacturers.

Further examples beyond the basic definitions in the main text of the standard are given in the National Appendix (NA) and are described in Form 1 to Form 4 below.

Form 1

This form provides for an enclosure to provide protection against direct contact with live parts (this is now 'basic protection') but does not provide any internal separation of switching, isolation or control items or terminations. These overall assemblies are often known as 'wardrobe' type with large front opening doors, usually with an integral door-interlocked isolator. Operating the isolator interrupts all functions but allows the door to be opened to gain access to the assembly for installation or maintenance. Such assemblies normally have lower fault withstand and it may be inconvenient to shut down a whole plant or system for a simple maintenance or repair operation (see Figure 2.1).

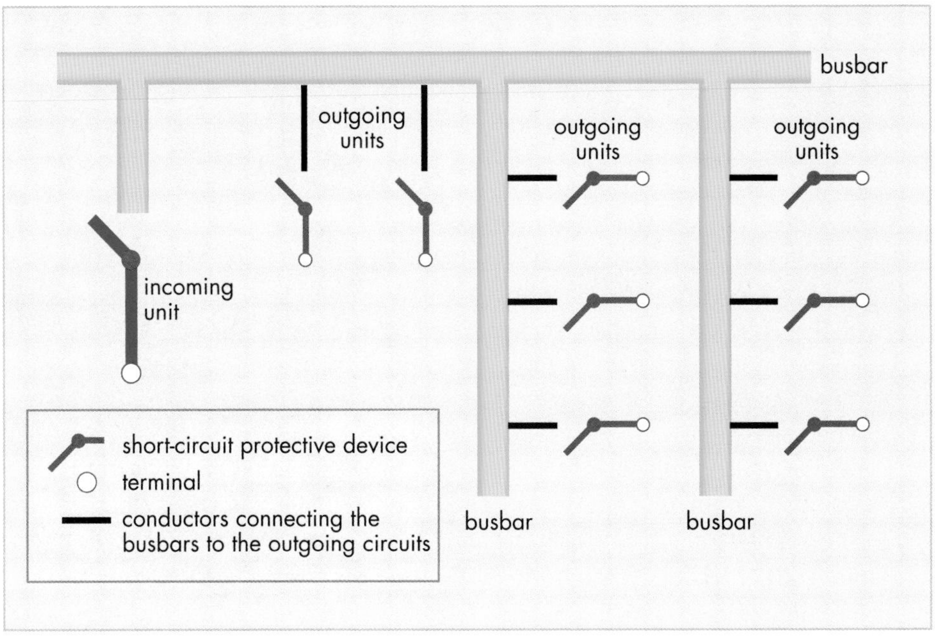

▼ **Figure 2.1**
Form 1 construction

Form 2

The overall assembly enclosure provides protection against direct contact with live parts (basic protection); separation is provided between the busbar assembly and switching, isolation, control items and terminations. There is very little advantage of this over Form 1 and the style is similar. Form 2 can be subdivided into:

- Type 1, in which the busbars are separated by insulation of the bars.
- Type 2, in which the busbars are separated by metallic or non-metallic rigid barriers.

Form 3

The enclosure provides protection against direct contact with live parts (basic protection) and also separation is provided between the busbars and switching, isolation or control items and between all these items. Outgoing terminals are not separated from each other, or perhaps from the busbars. Form 3 can be subdivided into:

- Form 3a, in which outgoing terminals are not separated from the busbars.
- Form 3b, in which outgoing terminals are separated from the busbars.

Form 3b can be further subdivided into:

- Type 1, in which the busbars are insulated for separation.
- Type 2, in which busbar separation is by metallic or non-metallic rigid barriers or partitions.

Form 4

The enclosure provides protection against direct contact with live parts (basic protection) and internal separation of the busbars from all switching, isolation and control items and outgoing terminations and separation of all items and outgoing terminations from each other. This allows for access to any single item, such as a switch-fuse or starter and its outgoing terminations, to enable work to be carried out whilst the assembly remains operational. Protection is also provided against an arcing fault in one device affecting other items. This is the usual form specified for commercial and industrial switchgear and controlgear assemblies but the designer has to consider whether, due to the extra cost, such requirements are necessary or justified. Form 4 can be subdivided into seven types:

- Type 1, in which the busbars are separated by insulation coverings. Terminals for external conductors are in the same compartment as the associated item of switchgear, etc., but cables may be glanded elsewhere.
- Type 2, in which the busbars are separated by metallic or non-metallic rigid barriers or partitions. Terminals or external conductors are in the same compartment as the associated item of switchgear, etc., but cables may be glanded elsewhere.
- Type 3, in which separation requirements are by metallic or non-metallic rigid barriers or partitions. Terminals or external conductors are in the same compartment as the associated item of switchgear, etc. and each item has its own integral cable glanding facility.
- Type 4, in which the busbars are separated by insulated coverings. Terminals for external conductors are not in the same compartment as the associated item of switchgear, etc., but in separate enclosed spaces. However, cables may be glanded elsewhere.
- Type 5, in which busbars are separated by metallic or non-metallic rigid barriers or partitions. Terminals for external conductors are not in the same compartment as the associated item of switchgear, etc., but in separate enclosed spaces and terminals may be separated by insulated coverings. Cables may be glanded in common cabling chambers.
- Type 6, in which all separation requirements are by metallic or non-metallic rigid barriers or partitions. Terminals for external conductors are not in the same compartment as the associated item of switchgear, etc., but in separate enclosed spaces and cables are glanded in common cabling chambers.
- Type 7, in which all separation requirements are by metallic or non-metallic rigid barriers or partitions. Terminals for external conductors are not in the same compartment as the associated item of switchgear, etc., but in separate enclosed spaces and the termination for each item has its own integral glanding facility (see Figure 2.2).

Switchboard manufacturers therefore cannot give all-embracing assurances for safe working, according to the form of separation with parts of the assembly energized. Specifying a particular form of separation will not guarantee this for any given form number.

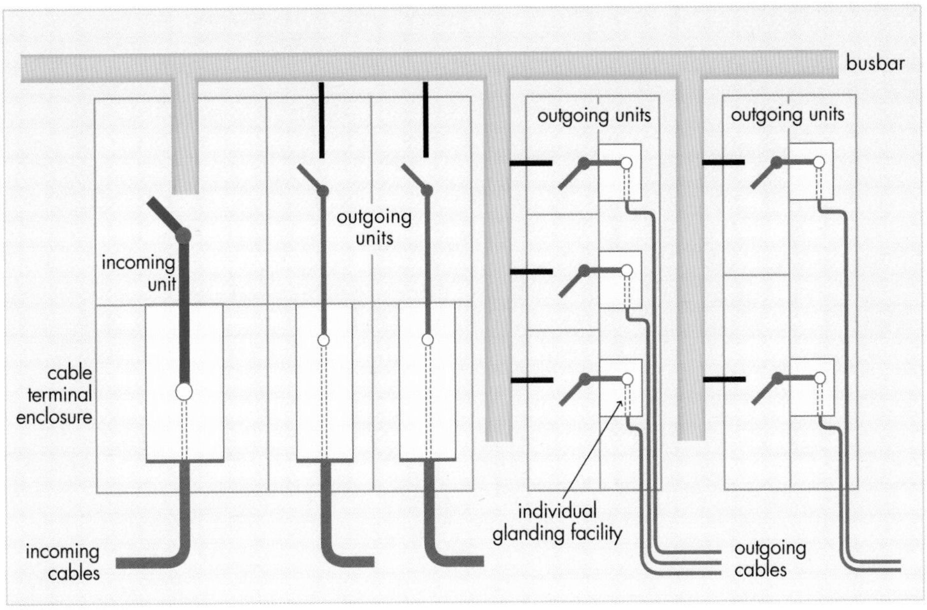

▼ **Figure 2.2**
Form 4 construction (Type 7)

2.8.2 General

Assemblies are to be designed and constructed so as to be able to withstand the thermal and dynamic stresses resulting from fault currents up to their rated values. The installation designer must specify the prospective fault current conditions at the point of installation.

Busbar systems for switchgear and controlgear should be adequately rated for the normal duty and maximum fault current level expected and should be well supported and braced, as the electromechanical stresses under fault conditions can be severe. However, thermal movement must be allowed for. It is usual for manufacturers to prove their busbar designs with a full short-circuit test by an independent test station up to the rated fault current level on a sample construction. The busbar arrangement can then be certified as a 'type-tested' busbar system.

Tests, as detailed in BS EN 60439-1, are applied to the complete assembly as relevant at manufacture, including continuity, insulation resistance and perhaps a flash test at high voltage. The designer and installer must be aware that there are statutory requirements under the Electricity at Work Regulations and the CDM Regulations, etc., for the safe design, construction, operation and provision for maintenance of electrical equipment assemblies. Adequate access, working space and lighting need to be provided where work is to be carried out on or near equipment, in order that persons may work safely.

It should also be realised that Regulation 8 of the Electricity at Work Regulations places an absolute requirement (one that shall be met regardless of cost or any other consideration) on protective conductor connections to earth:

> *... a conductor shall be regarded as earthed when it is connected to the general mass of earth by conductors of sufficient strength and current-carrying capability to discharge electrical energy to earth.*

It may be questioned whether the termination of steel or aluminium wire armour (where used as a protective conductor) with glands into metal gland plates, which themselves may only be bolted to the switchgear or controlgear frame, is an adequate connection. Cable gland 'earth tags' and supplementary connections to the equipment earth terminals may be necessary. In any case it should be ensured that any paint or other surface finish on the switchgear does not prevent effective electrical continuity between the adjacent parts.

612.13.2 During site installation and commissioning, tests as required by BS 7671 Part 6 should be carried out on the complete assembly, plus any other specific tests advised by the manufacturer or required by the client, user or engineer. It is not usual to carry out a repeat of specialist tests, e.g. a flash test at site; however, in the event of such a requirement or request, the manufacturer's advice should be sought.

Protection against overcurrent, electric shock and overvoltage

3

3.1 Protection against electric shock – overview

Chapter 41 of BS 7671 deals with protection against electric shock as applied to electrical installations. It is based on IEC 61140, which is a basic safety standard that applies to the protection of persons and livestock. BS EN 61140 is intended to give fundamental principles and requirements that are common to electrical installations and equipment or are necessary for their coordination. Section 410

The fundamental rule of protection against electric shock, according to BS EN 61140, is that:

- hazardous-live-parts shall not be accessible, and
- accessible conductive parts shall not be hazardous-live, either in use without a fault or in single-fault conditions.

According to 4.2 of BS EN 61140, protection under normal conditions is provided by **basic** protective provisions and protection under single-fault conditions is provided by **fault** protective provisions. Alternatively, protection against electric shock is provided by an **enhanced** protective provision which provides protection in use without a fault and under single-fault conditions.

Note that in BS 7671:2001, the 16th Edition of the *IEE Wiring Regulations*:

1 protection in use without a fault (now designated basic protection) was referred to as protection against direct contact, and
2 protection under fault conditions (now designated fault protection) was referred to as protection against indirect contact.

BS 7671:2008 generally recognises four methods of protecting against electric shock, which are to be applied in each part of an installation: 410.3.3

1 Automatic disconnection of supply
2 Double or reinforced insulation
3 Electrical separation for supply to one item of current-using equipment
4 Extra-low voltage (SELV and PELV).

See Guidance Note 5: *Protection Against Electric Shock*, for more detailed information on items 1 to 4 above. Item 1 is also described in more detail in this chapter.

411.1 ### 3.1.1 Automatic disconnection of supply

The protective measure of automatic disconnection of supply is employed in almost every electrical installation and consists of basic protection, fault protection and, for some circuits and locations, additional protection.

3.1.2 Basic protection

Basic protection is provided by basic insulation of live parts and/or by barriers or enclosures.

3.1.3 Fault protection

Fault protection is provided by protective earthing, protective equipotential bonding and automatic disconnection of supply in case of a fault.

This protective measure is intended to reduce the risk of shock by limiting the disconnection time and voltage magnitude in the event of a fault.

Where it is not possible to achieve a sufficiently low earth fault loop impedance to operate an overcurrent device within the required disconnection time, an RCD may be used to meet the requirements for fault protection.

411.3.2.6 Where automatic disconnection cannot be achieved in the required time, supplementary
415.2 equipotential bonding may also be provided for all or part of an installation in addition to, or instead of, an RCD. Low earth fault loop impedance values, however, lead to high fault current and faster disconnection times.

In TN systems, it is preferable for reliable operation for fault protection to be provided by overcurrent devices, including RCBOs operating as overcurrent devices.

During a fault, voltages are present along the fault path. The protection must ensure that any such voltages are not likely to persist long enough to be dangerous, as access to such voltages may result in electric shock. The protective bonding will have voltages present along it if it forms part of the fault path. Supplementary equipotential bonding may be required in order to keep the magnitude of the voltages between simultaneously accessible exposed-conductive-parts and/or extraneous-conductive-parts low. (See Part 7 of BS 7671 for specific requirements in special locations. Guidance Note 7 also addresses these requirements.)

Section 415 ### 3.1.4 Additional protection

Additional protection must be provided by:

411.3.3 ▶ one or more 30 mA residual current devices (RCDs) for specified circuits, and
415.2 ▶ protective supplementary equipotential bonding where required.

3.2 Protective devices

The Regulations prescribe requirements for the following devices:

1 Overcurrent protective devices
2 Residual current devices
3 Insulation monitoring devices.

3.3 Overcurrent protective devices

Every circuit must be provided with a means of overload and fault current protection although some parts of a circuit, such as the flexible cord of a luminaire pendant, may be protected against overload by the nature of the load. The devices may be either fuses or circuit-breakers. For TN systems such overcurrent devices can also be used for fault protection, provided that the earth fault loop impedance of the protected circuit is such that the chosen device will operate within the times specified in BS 7671. Regulation 411.5.2 (note 1) also refers to the use of the overcurrent protective device in a TT installation if the earth loop impedance is sufficiently low and is likely to remain so. However, RCDs are preferred and are usually specified. For TN systems, Tables 41.2, 41.3 and 41.4 of BS 7671 give maximum permitted earth fault loop impedances. For other types of device and current ratings the manufacturer should be consulted.

533.2
533.3
411.4.4
411.4.5
Table 41.1
Table 41.2
Table 41.3
Table 41.4
411.5.2

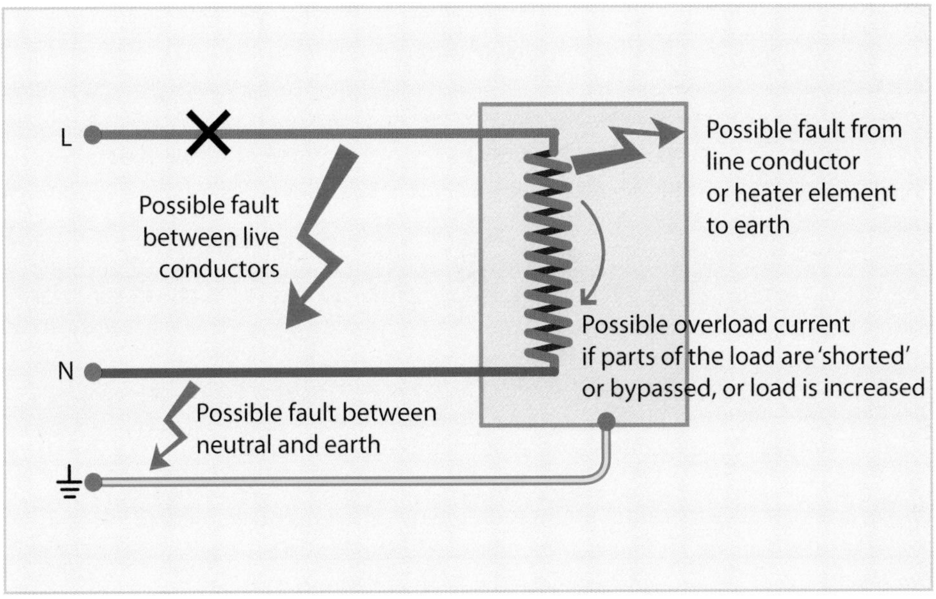

▼ **Figure 3.1**
How an overcurrent/fault current may occur in a single-phase circuit

The choice of protective device will depend on a number of factors, including overall installation and maintenance costs. Table 3.1 gives a brief comparison between fuses and circuit-breakers.

▼ **Table 3.1**
Comparison between fuses and circuit-breakers

Semi-enclosed fuses	Cartridge fuses	Circuit-breakers
Lowest capital cost	Higher capital cost	Highest capital cost
Low maintenance costs	Higher maintenance costs	No maintenance cost
Lowest fault current capacity	Highest fault current capacity	Intermediate fault current capacity
Requires a degree of skill to replace fusewire, fitting of wrong size/rating wire is not prevented	Relatively simple to replace cartridge. Size of cartridge relates to size of fuse carrier in consumer units for domestic installation	Easy to reset. Unskilled persons able to reset. No replacement or refitting after operation, unless damaged by operation at high fault currents. Can operate on surges.

3.3.1 Motor circuits

With respect to electric motors, note should be taken of BS EN 60947-4-1. This
536.5.2 recognises three types of coordination with corresponding levels of permissible damage to the starter. Coordination of Type 2 in BS EN 60947-4-1 (i.e. no damage, including permanent alteration of the characteristics of the overload relay, except light contact burning and a risk of contact welding) is required if the starter is to continue to provide
433.1 overload protection complying with Regulation 433.1. Examination and maintenance of the devices is always necessary after a fault.

The selected fuse or circuit-breaker provides short-circuit protection for the motor circuit and the starter overload relay provides protection from overload. In motor circuits the short-circuit protection and overload protection are often provided separately. Overload relays must be selected for the motor duty. For unusual duties, i.e. frequent starting/stopping, the manufacturer should be consulted.

The overload relay on the starter is arranged to operate for values of current from just above full load to the overload limit of the motor but it has a time delay such that it does not respond to either starting currents or fault currents. This delay provides discrimination with the characteristics of the associated fuse or circuit-breaker at the origin of the circuit on fault currents and prevents operation due to surge currents at start-up.

433.1 The starter overload relay can provide overload protection for the circuit in compliance with Regulation 433.1 on the basis that:

1 its nameplate full-load current rating or setting is taken as I_n, and
552.1.1 **2** the motor full-load current is taken as I_b, and
3 the ultimate tripping current of the overload relay is taken as I_2.

Where the overload relay has a range of settings then items (1) and (3) should be based on the highest current setting, unless the setting cannot be changed without the use of a tool. (See Guidance Note 6 for further information.)

3.4 Fuses

Fuses have a rated short-circuit capacity (see Table 3.2) and should be selected such that their rating is not exceeded by the prospective fault current at the point of installation, unless adequate back-up protection is specified.

▼ **Table 3.2**
Rated short-circuit capacities of fuses

Device type	Device designation	Rated short-circuit capacity (kA)
Semi-enclosed fuse to BS 3036 with category of duty	S1A S2A S4A	1 2 4
Cartridge fuse to BS 1361 type I type II		16.5 33.0
General purpose fuse to BS 88-2.2		16.5 at 240 V 80 at 415 V

Appx 3 BS 88-2.2 and BS 88-6 cartridge fuses have the same characteristics, as shown in Appendix 3 of BS 7671.

The size of tinned copper single-wire element for use where BS 3036 semi-enclosed (rewirable) fuses are selected and there are no manufacturer's instructions is given in Table 3.3. 533.1.1.3 Table 53.1

▼ **Table 3.3**
Sizes of tinned copper wire for use in semi-enclosed fuses (from Table 53.1 of BS 7671)

Rated current of fuse element (A)	Nominal diameter of wire (mm)
3	0.15
5	0.2
10	0.35
15	0.5
20	0.6
25	0.75
30	0.85
45	1.25
60	1.53
80	1.8
100	2.0

BS 3036 fuses have a limited fault current breaking capacity and also cannot be relied upon to operate within 4 hours at 1.45 times the nominal current of the fuse element. Correct protection can be obtained by modifying the normal condition $I_n \leq I_z$ such that the fuse rating does not exceed 1.45/2 = 0.725 times the rating of the circuit conductor. For this reason, larger cables may need to be selected where overload protection is provided by semi-enclosed fuses than when it is provided by a cartridge fuse or fuses or circuit-breaker. (Appendix 4 of BS 7671 and Guidance Note 6 give further guidance.) Appx 3 433.1.3 Appx 4 GN6

Note should be taken of the implied warning regarding the possible inadvertent replacement of a fuse link by one of a higher nominal current rating. 533.1.1.2

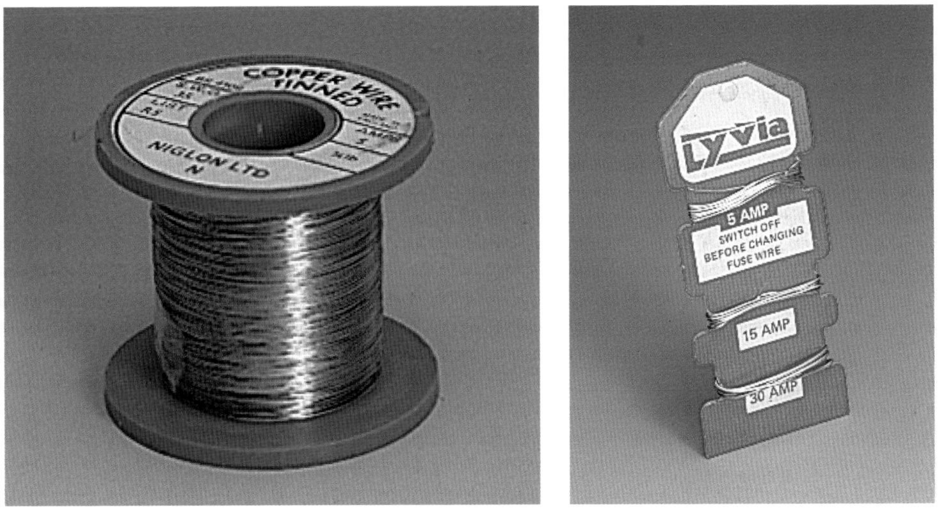

▼ **Figure 3.2**
Fusewire

3.5 Circuit-breakers

3.5.1 Circuit-breakers for applications not exceeding 440 V a.c.

There is a wide range of circuit-breaker characteristics that have been classified according to their instantaneous trip performance and Table 3.4 gives some information on the applications of the various types available. These limits are the maximum allowed in circuit-breakers to BS 3871 (now withdrawn) and circuit-breakers to BS EN 60898-1; note that manufacturers may provide closer limits. For such circuit-breakers, manufacturers' data for I_a may be applied to the formula given in Regulation 411.4.5. (I_a is the current causing automatic operation of the device within the stated time.) Appx 3 411.4.5

Table 3.4 Circuit-breakers overcurrent protection

Type	Multiple of rated current I_n below which it will not trip within 100 ms	Multiple of rated current I_n above which it will definitely trip within 100 ms	Typical application
1 B	2.7X 3X	4X 5X	Circuits not subject to inrush currents/ switching surges
2 C	4X 5X	7X 10X	Circuits where some inrush current may occur. For general purpose use on fluorescent lighting circuits, small motors, etc.
C 3	5X 7X	10X 10X	Circuits where high inrush currents are likely, e.g. motors, large lighting loads, large air conditioning units
D 4	10X 10X	20X 50X	Circuits where inrush currents are particularly severe, e.g. welding machines, x-ray machines

Notes:

1. Some non-linear resistive loads, such as large tungsten filament lighting installations, may give rise to high inrush currents.
2. Due to the withdrawal of BS 3871, Types 1 to 4 circuit-breakers are now obsolete but are still to be found in use.
3. Type D and Type 4 CBs are special-purpose circuit-breakers that require low earth fault loop impedance and should not be used without due consideration, particularly where this device is to meet the requirements for fault protection. Data should be obtained from the manufacturer.

3.5.2 Circuit-breakers for applications not exceeding 1000 V a.c. (MCCBs)

Unlike circuit-breakers to BS EN 60898-1, circuit-breakers complying with BS EN 60947-2 do not have defined characteristics and manufacturers' data must be used. BS EN 60947-2:2006 includes Annex L, which is for circuit-breakers not fulfilling the requirement for overcurrent protection (CBIs), derived from the equivalent circuit-breaker. A class X CBI is fitted with integral short-circuit protection, which may, on the basis of the manufacturer's data, be used in conjunction with the starter overload relay, for short-circuit protection.

3.5.3 Circuit-breakers – general

All circuit-breakers have a maximum fault current breaking capacity and care is needed in selection to ensure that this is not exceeded in service. BS 3871 identified this capacity with an 'M' rating; however, in BS EN 60898 the 'M' category ratings for breaking capacities are replaced by a figure in a rectangle (I_{cn}) on the device (see Table 3.5). The manufacturer has to declare the breaking capacity of the devices at specified power factors of test current. Higher fault current capacities up to 25 kA are recognised for BS EN 60898 devices. Rated values for both standards are given in Table 3.5.

▼ **Table 3.5**
Rated breaking capacity for miniature circuit-breakers to BS 3871, circuit-breakers to BS EN 60898 and RCBOs to BS EN 61009

Miniature CBs to BS 3871

'M' number marked on device	I_{cn} (A)
M1	1000
M1.5	1500
M3	3000
M4.5	4500
M6	6000
M9	9000

CBs to BS EN 60898 and RCBOs to BS EN 61009

Marked on device in rectangle	I_{cn} rated capacity (A)	k	I_{cs} service capacity (A)
1500	1500	1	1500
3000	3000	1	3000
4500	4500	1	4500
6000	6000	1	6000
10000	10000	0.75	7500
15000	15000	0.5	7500
20000	20000	0.5	10000
25000	25000	0.5	12500

Note: k is the ratio between service short-circuit capacity (I_{cs}) and rated short-circuit capacity (I_{cn}), BS EN 60898 and BS EN 61009.

Two short-circuit breaking capacities, I_{cn} and I_{cs}, are quoted for circuit-breakers:

I_{cn} the rated short-circuit capacity. This is the ultimate short-circuit breaking capacity.
I_{cs} the service short-circuit capacity. This is the maximum level of fault current operation after which further service is assumed without loss of performance.

For an assigned I_{cn} the I_{cs} value will not be less than the value tabulated in Table 3.5. The I_{cn} of the circuit-breaker must always exceed the prospective short-circuit current at the point of installation, except where back-up protection as specified by the manufacturer is applied.

Fault currents up to 16 kA

Except for London and some other major city centres, the maximum fault current for 230 V single-phase supplies up to 100 A will not exceed 16 kA. The short-circuit capacity of overcurrent protective devices incorporated within consumer units may be taken to be 16 kA where:

- the current ratings of the devices do not exceed 50 A
- the consumer unit complies with BS 5486-13 or BS EN 60439-3
- the consumer unit is supplied through a type 2 fuse to BS 1361:1971 rated at no more than 100 A.

Moulded case circuit-breakers

As mentioned in 3.5.2, moulded case and other types of circuit-breaker to BS EN 60947-2 do not have defined characteristics, unlike circuit-breakers to BS 3871 or BS EN 60898, and therefore manufacturers' data must be used.

3.6 Insulation monitoring devices

411.6.3 Devices which monitor and indicate the condition of the insulation must be installed in
IT systems to indicate a first fault to earth. Such systems may also be used where high
560.5.3 degrees of reliability of supply are necessary, as in supplies for safety services.

The first fault on such a system would then signal the need for remedial action, allowing time to carry this out before a possible second fault arose (see Figure 3.3).

Such a system calls for a special knowledge and is outside the scope of this Guidance
114.1 Note. An IT system is not permitted for low voltage public supply in the UK. The use
of IT systems is generally confined to industrial control systems, hospital operating
717.411.6 theatres and treatment rooms or outside broadcast units used for television.

▼ **Figure 3.3** Distribution network with insulation monitoring

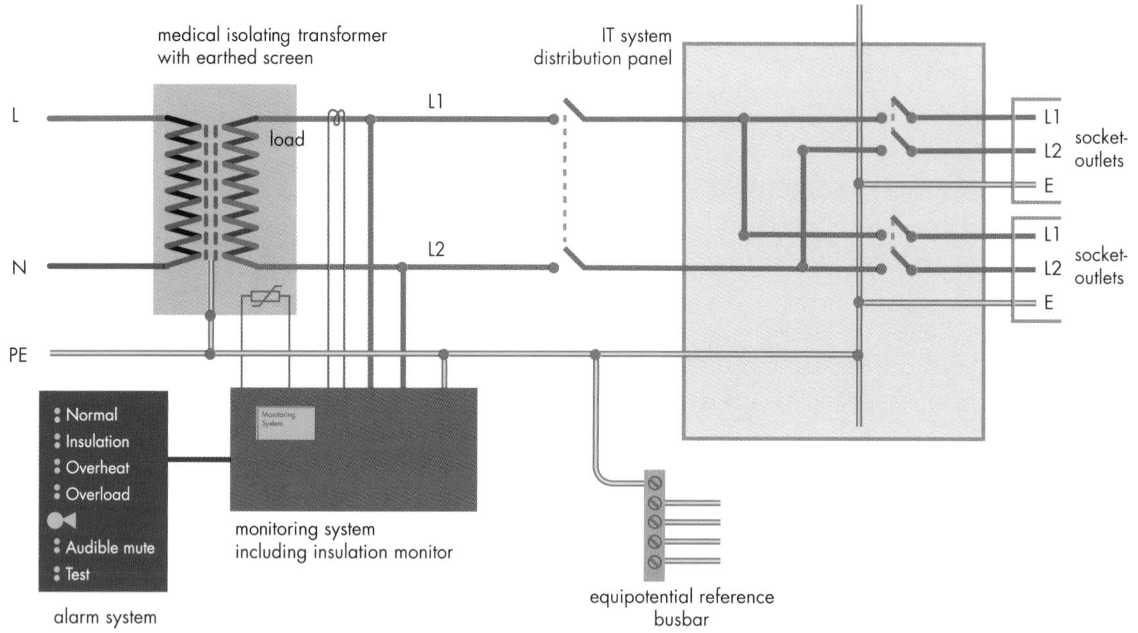

3.7 Residual current operated devices (RCDs)

Residual current device (RCD) is the generic term for a device that operates when the residual current in the circuit reaches a predetermined value. An RCD is a protective device used to automatically disconnect the electrical supply when an imbalance is detected between the line and neutral conductors. In the case of a single-phase circuit, the device monitors the difference in currents between the line and neutral conductors.

In a healthy circuit, where there is no earth fault current or protective conductor current, the sum of the currents in the line and neutral conductors is zero. If a line-to-earth fault develops, a portion of the line conductor current will not return through the neutral conductor. The device monitors this difference, and operates to disconnect the circuit when the residual current reaches a preset limit, the residual operating current ($I_{\Delta n}$).

Residual current protection may be provided by an assembly of detectors and relays, particularly in larger circuits. Where a power source is needed in order to operate the circuit-breaker concerned (known as voltage-dependent devices), the design must ensure that the operating supply is available at all times.

3.7.1 Terminology

There are a number of terms in use with RCD products, described below.

RCD (residual current device)
This is the generic term for all products which use the principle of detecting earth fault current by measuring the difference in current magnitude flowing in different supply conductors.

RCCB (residual current operated circuit-breaker without integral overcurrent protection)
A mechanical switching device designed to make, carry and break currents under normal service conditions and to cause the opening of the contacts when the residual current attains a given value under specific conditions. It is not designed to perform the functions of giving protection against overload and/or short-circuit and must always be used in conjunction with a circuit protective device.

RCBO (residual current operated circuit-breaker (RCCB) with integral overcurrent protection)
A mechanical switching device designed to make, carry and break currents under normal service conditions and to cause the opening of the contacts when the residual current attains a given value under specific conditions. In addition, it is designed to perform the functions of giving protection against overload and/or short-circuit and can be used independently of any other circuit protective device within its rated short-circuit capacity.

CBR (circuit-breaker incorporating residual current protection)
A circuit-breaker providing overcurrent protection and incorporating residual current protection either integrally (an integral CBR) or by combination with a residual current unit which may be factory or field fitted.

Note: The RCBO and CBR have the same application, both providing overcurrent and residual current protection. In general, the term RCBO is applied to the smaller devices whereas CBR is used for devices throughout the current range, with ratings up to several thousand amperes, single- and multi-phase. RCBO and CBR are more strictly defined by the relevant standards.

SRCD (socket-outlet incorporating a residual current device)
A socket-outlet for fixed installations incorporating an integral sensing circuit that will automatically cause the switching contacts in the main circuit to open at a predetermined value of residual current.

PRCD (portable residual current device)
A device consisting of a plug, usually to BS 1363, 13 A type, a residual current device and one or more socket-outlets (or a provision for connection). It may incorporate overcurrent protection.

RCM (residual current monitor)
A device designed to monitor electrical installations or circuits for the presence of unbalanced earth fault currents. It does not incorporate any tripping device or overcurrent protection.

MRCD (modular residual current device)
An independently mounted device incorporating residual current protection, without overcurrent protection and capable of giving a signal to trip an associated switching device.

SRCBOs (socket-outlets incorporating RCBOs)
These devices are RCBOs – RCCBs with integral overcurrent protection – connected to fixed socket-outlets.

Table 3.6
Types of RCD with relevant standard number and description

Device	Standard	Title
RCBO	BS EN 61009-1:2004	Residual current operated circuit-breakers with integral overcurrent protection for household and similar uses (RCBOs)
RCCB	BS EN 61008-1:2004	Residual current operated circuit-breakers without integral overcurrent protection for household and similar uses (RCCBs)
CBR	BS EN 60947-2:2006	Low-voltage switchgear and controlgear – Part 2: Circuit-breakers
SRCD	BS 7288:1990	Specification for socket-outlets incorporating residual current devices (SRCDs)
PRCD	BS 7071:1992	Specification for portable residual current devices
MRCD	BS EN 60947-2:2006	Low-voltage switchgear and controlgear – Part 2: Circuit-breakers
RCM	BS EN 62020:1999	Electrical accessories – Residual current monitors for household and similar uses (RCMs)

International and European standards are being prepared for SRCDs and PRCDs. In addition, a European standard for SRCBOs is being prepared.

RCDs for load currents below 100 A usually include the transformer and contact system within the same enclosure. Devices for load currents greater than 100 A usually comprise a transformer assembly with a detector and a separate shunt trip circuit-breaker unit or contactor, mounted together.

3.7.2 A wide choice of device

When selecting RCDs, the user is provided with a choice of residual current sensitivity. Typically, values of residual operating current ($I_{\Delta n}$) are between 10 mA and 2 A, with
531.2.10 some devices offering adjustment of sensitivity. Adjustable devices must not be installed where they would be accessible to unauthorized persons. It is, however, normal to expect that such devices will be fitted with either a cover that can be sealed or a cover that requires a tool to gain access to the means of adjustment.

As shown in section 3.7.1, devices are available with or without integral overcurrent protection or may be associated with an independent modular residual current relay.

The operating function of RCDs may be either independent or dependent on line
531.2.6 voltage/dependent on an auxiliary supply. BS 7671 states that RCDs requiring an auxiliary supply and which do not operate automatically in the case of failure of that source shall be used only if:

1. fault protection is maintained by other means, or
2. the device is in an installation operated and maintained by an instructed person or skilled person (because such persons should be aware of the risk).

This requirement reflects the risk of failure of the auxiliary source which could occur whilst the supply voltage remains present. In these circumstances the RCD may not operate although the circuit supply is healthy and the user may not be aware of the auxiliary power supply failure.

A test device is incorporated to allow the operation of the RCD to be checked. Operation of this device creates an out-of-balance condition within the device which establishes the integrity of the electrical and mechanical elements of the tripping device only. It should be noted that the test device does not provide a means of checking the continuity of any part of the earth path nor does it check the minimum operating current or operating time of the RCD.

The introduction of BS EN 61008 and BS EN 61009 extended the classification of residual current devices beyond that of the old UK standard BS 4293 and they include classification of such attributes as time delay facilities and operating characteristics for currents with d.c. components. RCDs are categorised into three types as detailed below, each of which is available in the 'General' type and the 'S' (Selective) type which incorporates a time delay to provide discrimination of RCDs when connected in series (see 3.7.3). An RCD used to meet the requirements for basic protection must be of the 'General' type (without intentional time delay). In a TN system, the time delay of the 'S' type RCD must not exceed the requirements of Regulation 411.4. 411.4

Types 'AC', 'A' and 'B' RCDs

- Type 'AC' RCDs will provide protection against a.c. earth fault currents
- Type 'A' RCDs will provide protection against a.c. earth fault currents that contain pulsating d.c. components
- Type 'B' RCDs will provide protection against d.c earth fault currents.

For the majority of applications, type AC devices are suitable, with type A being used where special circumstances exist. Type B devices are manufactured and tested to IEC 60755, which is a Technical Report, with no BS or BS EN equivalent. There is no requirement for the use of type A RCDs in BS 7671 but type B may be required in a solar photovoltaic (PV) system. 712.411.3.2.1.2

Type A devices are important due to the wide range of electronic equipment available now which may produce a protective conductor current with a pulsating d.c. component and harmonic currents. These RCDs may utilize electronic circuitry or low remanence core material to more closely match the predetermined current/time trip characteristics allowed by the standards to achieve the most suitable performance characteristics.

3.7.3 Discrimination

Where, in order to minimize the risk of danger and inconvenience to the user, discrimination is required between residual current devices installed in series, the device operating characteristics must provide the required discrimination. A time delay should be provided in the 'upstream' device by the use of a type S device (see Figure 3.4). 531.2.9

Type S devices are identified by the symbol **S** marked on the device.

Tables 3.7 and 3.8 (extracted from BS EN 61008 and BS EN 61009) provide comparative data for the devices.

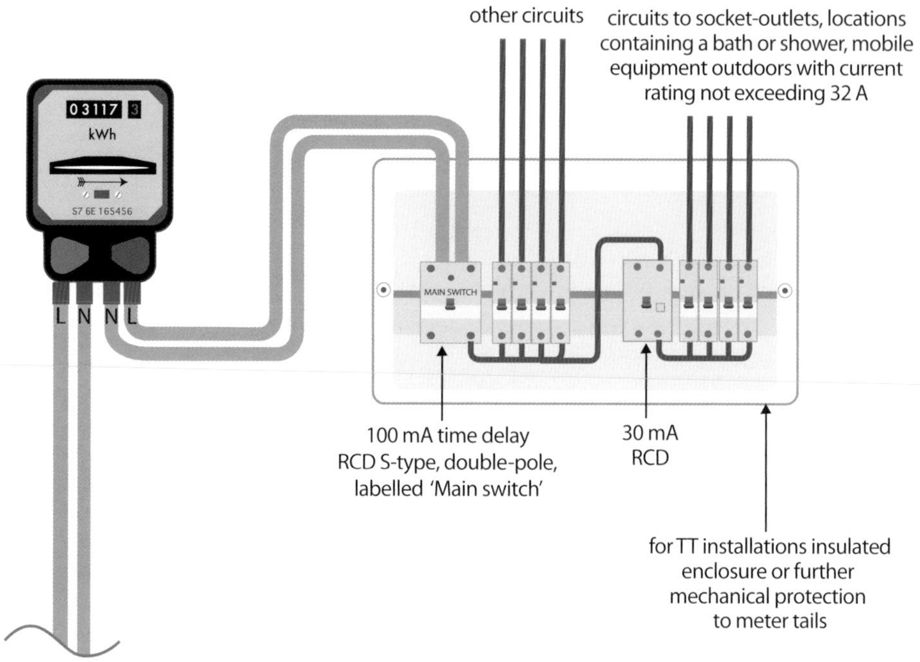

▼ Figure 3.4
Typical split consumer unit with time-delayed RCD as main switch, suitable for TT and TN installations with cables in walls or partitions having an earthed metallic covering or enclosed in earthed steel conduit or the like

▼ Table 3.7
Standard values of break time and non-actuating time for type AC RCCBs (extract from BS EN 61008)

Type	I_n	$I_{\Delta n}$	Standard values of break time (s) and non-actuating time (s) at a residual current (I_Δ) equal to:				
	A	A	$I_{\Delta n}$	$2\,I_{\Delta n}$	$5\,I_{\Delta n}$	500 A	
General	Any value	Any value	0.3	0.15	0.04	0.04	Maximum break times
S	≥ 25	> 0.030	0.5	0.2	0.15	0.15	Maximum break times
			0.13	0.06	0.05	0.04	Minimum non-actuating times

▼ Table 3.8
Standard values of break time and non-actuating time for operating under residual conditions for type AC RCBOs (extract from BS EN 61009)

Type	I_n	$I_{\Delta n}$	Standard values of break time (s) and non-actuating time (s) at a residual current (I_Δ) equal to:				
	A	A	$I_{\Delta n}$	$2\,I_{\Delta n}$	$5\,I_{\Delta n}$	$I_{\Delta t}$	
General	Any value	Any value	0.3	0.15	0.04	0.04	Maximum break times
S	≥ 25	> 0.030	0.5	0.2	0.15	0.15	Maximum break times
			0.13	0.06	0.05	0.04	Minimum non-actuating times

Definitions:

I_n = rated current of the device.
$I_{\Delta n}$ = rated residual operating current of the device.
$I_{\Delta t}$ = value of current ensuring residual current sensor does not operate before the overload sensor.

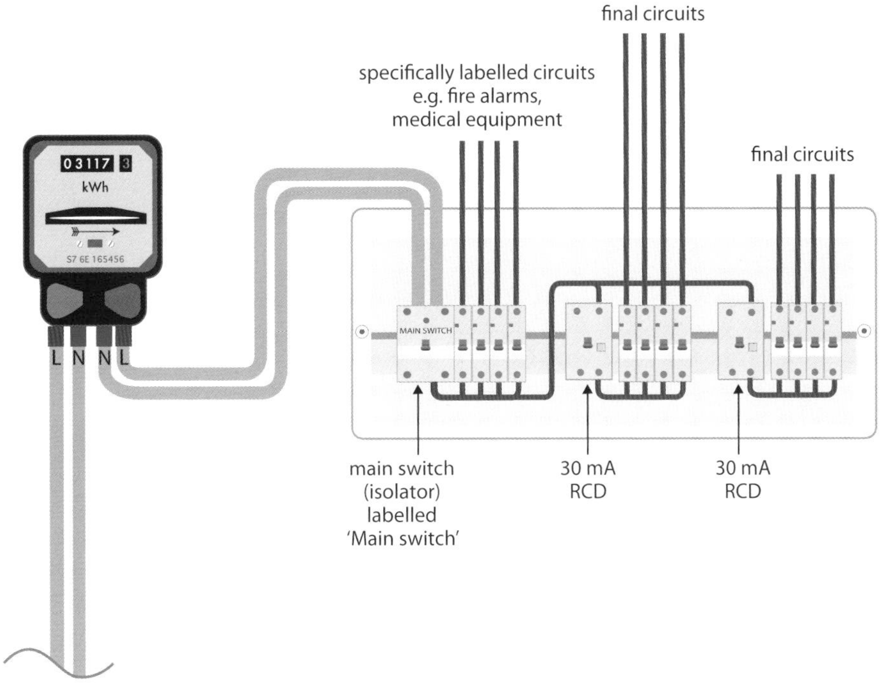

▼ **Figure 3.5** Three-way split consumer unit with separate main switch, two 30 mA RCDs and circuits without RCD protection

The purpose of an RCD in any installation is to provide protection against electric shock; however, it may be appropriate to exclude certain circuits, an example of which is shown in Figure 3.5.

RCDs with a suitable differential current rating and operating time are therefore suitable for:

- fault protection 411.1
- basic protection 415.1
- some protection against the risk of fire caused by fault currents. 422.3.9

RCDs are **intended to operate** to provide protection against line-to-earth and neutral-to-earth faults but **do not operate** in the event of line-to-line or line-to-neutral faults. 705.422.7

An RCD will only restrict the time during which a fault current flows. It cannot restrict the magnitude of the fault current, which depends solely on the circuit conditions. 415.1

For an installation which is part of a TN system the use of an RCD may be from choice or because the earth fault loop impedance is too high for overcurrent protective devices to operate within the maximum permitted disconnection times. In the latter case, a further option is to use supplementary equipotential bonding so that there will be no dangerous potential differences between simultaneously accessible exposed- and extraneous-conductive-parts. 411.4.4 411.3.2.6

3.7.4 Additional protection

BS 7671 requires that an RCD be used as additional protection for:

1 socket-outlets with a rated current not exceeding 20 A which are for use by ordinary persons and are intended for general use, and 411.3.3
2 mobile equipment with a current rating not exceeding 32 A for use outdoors.

As indicated in the Regulations, the RCD must have a rated residual operating current not exceeding 30 mA and an operating time not exceeding 40 ms at a residual current of 5 $I_{\Delta n}$. This requirement is met by general type devices complying with BS EN 61008 or BS EN 61009 (or BS 4293). 415.1

Exceptions are permitted where socket-outlets are used under the supervision of skilled or instructed persons, e.g. in some commercial or industrial locations, or where a specific labelled or otherwise suitably identified socket-outlet is provided for connection of a particular item of equipment, e.g. refrigeration or computing equipment.

3.7.5 Location of RCDs

RCDs must be part of the fixed installation and in domestic or residential installations can be installed at the consumer unit as either a separate RCD on all socket-outlet circuits (i.e. a split-load board) or a combined RCD/CB device (RCBO).

3.7.6 Selection of types of RCD

It is not good practice to provide a single RCD on the supply to a complete distribution board/consumer unit, which may have lighting circuits, etc., as operation will interrupt
314.1 the whole building supply. Regulation 314.1(i) requires installation circuit arrangements to be such that hazards are avoided and inconvenience minimized in the event of a fault. It can be argued that protection of an installation by an RCD is not related to the
531.4.1 'circuit' definition in Part 2 of BS 7671 and Regulation 531.4.1 makes specific reference to a single RCD for a TT installation. However, the decision to provide a single RCD, to use a split distribution board/consumer unit or to use RCBOs must be taken only after due consideration of all the relevant factors, including convenience and maintenance, and not just initial costs. For example, elderly or infirm persons may not be able to gain access to, or reset, an RCD. A single 30 mA RCD should normally not be installed at the origin of the installation.

3.7.7 TT systems

411.5.2 BS 7671 recognises that, in a TT system, either RCDs or overcurrent protective devices
411.5.3 may be used to provide fault protection by automatic disconnection of supply. In many such installations it will not be practicable to attain a sufficiently low value of earth electrode resistance to use overcurrent protective devices.

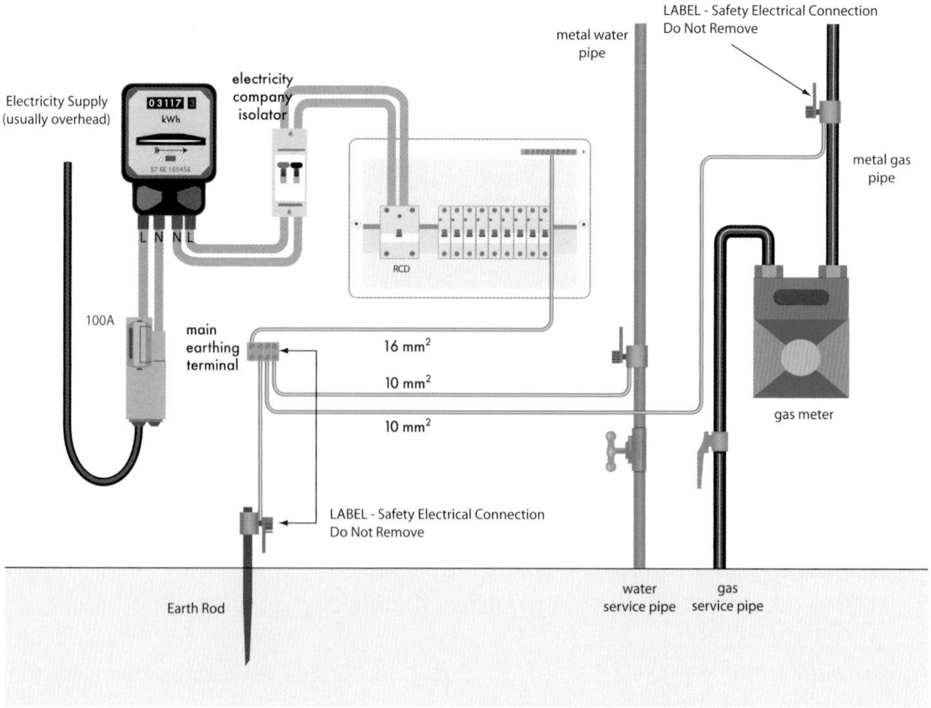

▼ Figure 3.6
Installation forming part of a TT system

Where an RCD is used (see Figure 3.6), the product of the rated residual operating current ($I_{\Delta n}$), in amperes, and the sum of the earth electrode and protective conductor resistances (R_A), in ohms, must not exceed 50 V. This does not mean, however, that the voltage to earth will not exceed 50 V or such other prescribed voltage; its purpose is to ensure that when the voltage does exceed the prescribed voltage the RCD will operate. In any event, to ensure compliance with the requirements of Regulation 411.5.3 the earth fault loop impedance must not exceed the appropriate limit given in Table 41.5 of BS 7671. 411.5.3 Table 41.5

3.7.8 Two distinct categories

From the foregoing, it is seen that it is possible to categorise RCDs into two distinct categories according to their fault current operating characteristics:

1. RCDs having a rated residual operating current greater than 30 mA. This group is intended solely to provide fault protection or protection against fire. 100 mA and 300 mA RCDs are sometimes said to protect against fire in the event of earth leakage. It should be noted that these units do not meet the requirements for additional protection afforded by 30 mA devices. 422.3.9 705.422.7
2. RCDs having a rated residual operating current of 30 mA or less and operating time not exceeding 40 ms at a residual current of 5 $I_{\Delta n}$ as provided by BS 4293, BS 7071, BS 7288, BS EN 61008-1 or BS EN 61009-1. This group is generally referred to as 'high sensitivity'. In addition to meeting the requirements for fault protection, they are recognised by BS 7671 as meeting the requirements for additional protection, that is, protection of persons who come into simultaneous contact with a live part and earth. 415.1

3.7.9 Limitations and precautions

To meet the requirements for basic protection incorporating the use of RCDs providing additional protection, it is essential that further measures are also applied (e.g. insulation). The use of an RCD in a circuit normally expected to have a protective conductor is not considered sufficient for protection for that circuit to meet the requirements for fault protection where there is no such protective conductor, even though the rated residual operating current of the RCD does not exceed 30 mA. 415.1 Section 416 531.2.5

As well as providing fault protection and additional protection, with a suitable differential current rating RCDs may provide some protection against fire risk, as stated earlier. The level of protection is related to the sensitivity of the device. For this purpose, an RCD should be chosen with the lowest suitable rated residual operating current. A lower operating current would give a greater degree of protection but it may also result in unwanted tripping. The connection of a further load at a later date may have an exacerbating effect due to increased leakage current.

The installation designer will often not know the sum total of the protective conductor currents occasioned by the loads. Neither will it always be known which equipment is going to be used or (if a number of circuits are to be protected by one RCD) how many of those items of equipment would be energized at any one time. 531.2.4 543.7

Knowing the use to which the installation will be put, the designer must deduce the likely total protective conductor current in the protected circuit or make an assessment and state this in the design. In cases of difficulty, circuits may be subdivided to reduce the possibility of unwanted tripping.

The total leakage of the various items of equipment protected by the RCD concerned should be such that any protective conductor current expected in normal service will not cause unwanted operation of the device. 531.2.4 543.7

British Standards with safety requirements for electrical equipment generally include limiting values of protective conductor current. Limits apply when cold and also at operating temperature. For instance, BS EN 60335-1 and BS 3456 Part 101 (which covers the general requirements for safety of household and similar electrical appliances) and other British Standards prescribe the limits shown in Appendix J.

3.7.10 Unwanted tripping

314.1 (iv) 531.2.4 Unwanted tripping (sometimes called 'nuisance tripping') of RCDs occurs when a leakage current causes unnecessary operation of the RCD. Such tripping may occur on heating elements, cooking appliances, etc., due to absorption of a small amount of moisture through imperfect element end seals when cold. Once energized, the moisture provides a conductive path for increased leakage and could cause the RCD to operate. The moisture dries out as the element heats up. Heating or cooking appliances are not, per se, required to be protected by RCDs. If the heating or cooking appliance 411.3.3 is supplied by a socket-outlet which is provided for use by ordinary persons then Regulation 411.3.3 requires that RCD protection at 30 mA is provided.

Further information on fault protection and application of RCDs is provided in Guidance Note 5. Details of the testing of RCDs are given in Guidance Note 3.

3.8 Automatic disconnection of supply

411.4 The tables of limiting impedances given in Chapter 41 of BS 7671 are for circuits having a nominal line voltage to Earth (U_0) of 230 V a.c. rms. For other values of U_0 the tabulated earth fault loop impedance values should be multiplied by $U_0/230$.

Appx 3 In Appendix 3 of BS 7671, curves have been drawn to represent the slowest circuit-breaker instantaneous operating times allowed by the circuit-breaker standards and median times for fuses. To assist the designer, a set of time/current values for specific operating times has been agreed for each device and is reproduced in a box at the right-hand side of each set of curves.

Regulations 411.3.2.2 to 411.3.2.4 specify maximum disconnection times for circuits:

- **Disconnection times for final circuits up to 32 A**
  411.3.2.2 Table 41.1 Regulation 411.3.2.2 requires that for a.c. circuits, where U_0 is between 120 V and 230 V, the disconnection time for final circuits up to 32 A in TN systems should not exceed 0.4 s; for TT systems this figure is reduced to 0.2 s.
- **Disconnection times for distribution circuits and for final circuits over 32 A**
  411.3.2.3 411.3.2.4 Distribution circuits, etc. in a TN system should disconnect within 5 s, and in a TT system they should disconnect within 1 s.

411.4 Regulations 411.4.6, 411.4.7 and 411.4.8 provide maximum earth fault loop impedances (Z_S) which will result in protective devices operating within the specified disconnection times (of Regulations 411.3.2.2 to 411.3.2.4).

411.4.5 The maximum earth fault loop impedance (Z_S) for a protective device is given by:

$$Z_S \times I_a \leq U_0$$

where:

Z_s is the impedance in ohms (Ω) of the fault loop comprising
- the source,
- the line conductor up to the point of the fault, and
- the protective conductor between the point of the fault and the source.

I_a is the current in amperes (A) causing the automatic operation of the disconnecting device within the time specified in Table 41.1 of Regulation 411.3.2.2 or, as appropriate, Regulation 411.3.2.3. Where an RCD is used, this current is the rated residual operating current providing disconnection in the time specified in Table 41.1 or Regulation 411.3.2.3.

U_0 is the nominal a.c. rms or d.c. line voltage to Earth in volts (V).

As stated in Appendix 3 of BS 7671, in Tables 41.2, 41.3 and 41.4 the nominal supply voltage U_0 is taken as 230 V.

The tabulated values in BS 7671 are generally applicable for nominal 230 V supplies within the statutory limits from Regional Electricity Companies. For other supplies the designer will need to determine the nominal voltage of the system and calculate Z_s accordingly.

Where it is not possible to obtain a sufficiently low earth fault loop impedance, one recognised solution would be to install local supplementary equipotential bonding. 411.3.2.6

Where supplementary equipotential bonding is to be installed, it is necessary to connect together the exposed-conductive-parts of equipment in the circuits concerned including the earthing contacts of socket-outlets and extraneous-conductive-parts. Supplementary bonding conductors must be selected to comply with the minimum size requirements of Regulation 544.2, and the resistance (R) of the supplementary bonding conductor between simultaneously accessible exposed-conductive-parts and extraneous-conductive-parts must fulfil the following condition: 415.2.1 544.2 415.2.2

$$R \leq \frac{50}{I_a}$$

where:

I_a is the operating current of the protective device:
- for an RCD, the rated residual operating current $I_{\Delta n}$ in amperes is used
- for an overcurrent device, the minimum current that disconnects the circuit within 5 s is used.

However, if local supplementary bonding is provided to limit shock voltage magnitude, there is still a requirement to disconnect the supply to the circuit for protection against thermal effects (see Appendix D). The circuit must be designed such that cables and equipment will not be damaged by the thermal effects of the fault current, whatever the disconnection time, as BS 7671 does not limit the disconnection time in this regard.

The application of RCDs is discussed in section 3.7.

Where, in the event of a fault, for reasons other than electric shock, the disconnection times of Table 41.1 or Regulations 411.3.2.3 or 411.3.2.4 cannot be met, the output voltage of the source must be reduced to 50 V a.c. or 120 V d.c. in the times stated. 411.3.2.5

3.8.1 Types of protective conductor

543.2.2 It is worth noting the provisions of Regulation 543.2.2 and especially that steel wire armour, metal conduit and trunking, etc., properly installed and maintained, are suitable for use as protective conductors where they can fulfil the shock, thermal protection requirements and minimum requirements for cross-sectional area of Section 543 of BS 7671. Separate copper protective conductors are *not* always required. However, for functional reasons, a functional earth connection may need a copper conductor. The manufacturer of the equipment should be consulted.

543.2.6 Extraneous-conductive-parts, such as structural steel or significant internal metal components, may be used as protective bonding conductors, provided they are electrically continuous, suitably sized and precautions are taken to prevent their removal.

3.8.2 Protective bonding

411.3.1.2 Main protective bonding conductors are required to connect extraneous-conductive-parts, including main services and metallic structural parts, as stated in Regulation 411.3.1.2, to the main earthing terminal. Conductors should be selected and installed
544.1 in accordance with the requirements of Regulation 544.1. Section 3.8.3 below gives guidance on required conductor sizes. Figure 3.7 gives examples of protective bonding arrangements.

Supplementary equipotential bonding is installed, as appropriate, in accordance with
544.2 the requirements of Regulation 544.2. Table 3.9 gives guidance on conductor sizes.

▼ Figure 3.7
Examples of protective bonding

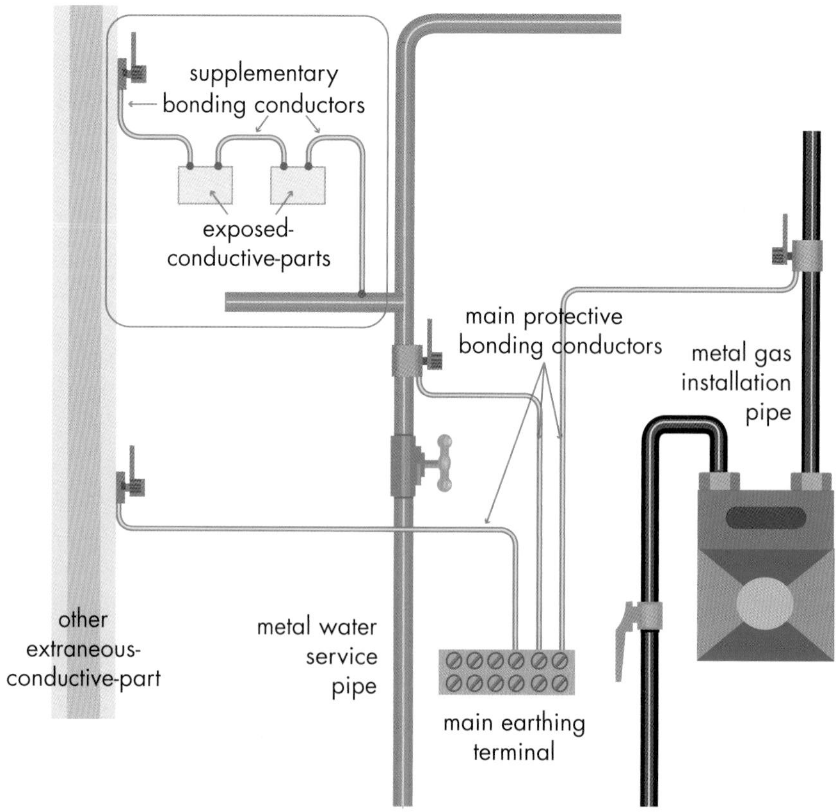

Size of circuit protective conductor (mm²)	Minimum cross-sectional area of supplementary bonding (mm²)					
	Exposed-conductive-part to extraneous-conductive-part		Exposed-conductive-part to exposed-conductive-part		Extraneous-conductive-part to extraneous-conductive-part*	
	mechanically protected	not mechanically protected	mechanically protected	not mechanically protected	mechanically protected	not mechanically protected
	1	2	3	4	5	6
1.0	1.0	4.0	1.0	4.0	2.5	4.0
1.5	1.0	4.0	1.5	4.0	2.5	4.0
2.5	1.5	4.0	2.5	4.0	2.5	4.0
4.0	2.5	4.0	4.0	4.0	2.5	4.0
6.0	4.0	4.0	6.0	6.0	2.5	4.0
10.0	6.0	6.0	10.0	10.0	2.5	4.0
16.0	10.0	10.0	16.0	16.0	2.5	4.0

* If one of the extraneous-conductive-parts is connected to an exposed-conductive-part, the bond must be no smaller than that required for bonds between exposed-conductive-parts and extraneous-conductive-parts – column 1 or 2.

▼ **Table 3.9**
Supplementary bonding conductors

Protective multiple earthing (PME) is now widely used for low voltage supplies (a TN-C-S system) and has specific requirements for earthing and bonding of the installation. These are discussed in more detail in Guidance Notes 5 and 8.

3.8.3 Sizing of earthing conductor and main protective bonding conductors

Csa of associated line conductor[1]	mm²	4	6	10	16	25	35	50	70
Csa of non-buried[2] earthing conductor[3]	mm²	4	6	10	16	16	16	25	35
Csa of main protective bonding conductors[4]	mm²	6	6	6	10	10	10	16	25

▼ **Table 3.10a**
Minimum cross-sectional areas (csa) of earthing conductor and main protective bonding conductors for TN-S systems

Notes:

1 The line conductor is the conductor between the origin of the installation and the consumer unit or main distribution board.

2 In the event that the earthing conductor is buried (see Table 54.1), it must have a csa of at least:

- 25 mm² copper or 50 mm² steel if not protected against corrosion by a sheath or protected against mechanical damage
- 16 mm² copper or 16 mm² coated steel if protected against corrosion by a sheath but not protected against mechanical damage.

3 The csa of the earthing conductor has been selected in accordance with Table 54.7 of Regulation 543.1.4. Note that calculation in accordance with Regulation 543.1.3 is also permitted.

4 The csa of the main protective bonding conductors is in accordance with Regulation 544.1.1.

Where application of the regulations produces a non-standard size, a conductor having the nearest larger standard csa has been used in Table 3.10a.

Table 3.10a applies for copper conductors. Copper equivalent sizes may be used.

Regional Electricity Companies may require conductors with a larger csa than those given above.

▼ Table 3.10b
Minimum cross-sectional areas (csa) of earthing conductor and main protective bonding conductors for TN-C-S (PME) systems

Csa of supply neutral conductor	mm^2	4	6	10	16	25	35	50	70
Csa of non-buried[1] earthing conductor[2]	mm^2	10	10	10	16	16	16	25	35
Csa of main protective bonding conductors[3]	mm^2	10	10	10	10	10	10	16	25

Notes:

1 In the event that the earthing conductor is buried (see Table 54.1), it must have a csa of at least:

- 25 mm^2 copper or 50 mm^2 steel if not protected against corrosion by a sheath or protected against mechanical damage
- 16 mm^2 copper or 16 mm^2 coated steel if protected against corrosion by a sheath but not protected against mechanical damage.

2 PME conditions apply, hence the csa of the earthing conductor must comply with the requirements of Sections 543 and 544. In the above table, the csa of the earthing conductor has been selected in accordance with Table 54.7 of Regulation 543.1.4 and, where applicable, Table 54.8 of Regulation 544.1.1.

3 The csa of the main protective bonding conductors is in accordance with Table 54.8.

Where application of the regulations produces a non-standard size, a conductor having the nearest larger standard csa has been used in Table 3.10b.

Table 3.10b applies for copper conductors. Copper equivalent sizes may be used.

Regional Electricity Companies may require conductors with a larger csa than those given above.

544.1.1 Except for highway power supplies and street furniture, where PME conditions apply the main protective bonding conductor must be selected in accordance with the neutral conductor of the supply and Table 54.8.

Where an installation has more than one source of supply to which PME conditions apply, a main protective bonding conductor must be selected according to the largest neutral conductor of the supplies.

3.8.4 Earthing of flush metal accessory boxes

It has been questioned whether flush metal accessory boxes come within the definition of 'exposed-conductive-parts' that require to be connected to the circuit protective conductor.

An exposed-conductive-part is defined as follows:

Part 2 *Conductive part of equipment which can be touched and which is not normally live, but which can become live when basic insulation fails.*

In normal use, such accessory boxes cannot be touched. Only when an accessory is removed can its metal accessory box be touched.

However, the view of the IET Wiring Regulations Policy Committee is that flush metal accessory boxes should be considered to be exposed-conductive-parts and connected to the main earthing terminal by means of a circuit protective conductor.

Where an accessory does not have an earthing terminal incorporated, such as certain light switches, the circuit protective conductor should be terminated at the earthing terminal in the accessory box.

Where an accessory has an earthing terminal incorporated, the protective conductor should be connected at the earthing terminal. Accessories which have an earthing terminal incorporated, such as socket-outlets, normally have an earth strap connecting the earthing terminal to one or both of the fixing holes. Flush metal accessory boxes usually have at least one fixed lug (see Figure 7.8). The IET Wiring Regulations Policy Committee has agreed that such boxes may be earthed via the tight metal-to-metal contact of the fixing screw in the fixed lug. Consequently, unless both lugs are of the adjustable type, it is not necessary to connect the earthing terminal of the accessory to the earthing terminal in the associated flush metal accessory box by a separate protective conductor (i.e. an earthing tail).

Where the circuit protective conductor is formed by conduit, trunking, ducting or the metal sheath and/or armour of a cable, an earthing tail must be fitted. 543.2.7

▼ **Figure 3.8**
Flush metal accessory box with one fixed lug and one adjustable lug

Where there is any doubt about the continued effectiveness, reliability and permanence of an earthing connection formed by the screws between an accessory and a metal box, it is recommended that a suitably sized earthing tail is provided.

3.9 Protection against voltage disturbances

Section 442 Chapter 44 of BS 7671 has now been expanded to include Section 442, *Protection of low voltage installations against temporary overvoltages due to earth faults in the high voltage system and due to faults in the low voltage system* joining the existing Section 443, *Protection against overvoltages of atmospheric origin or due to switching.* It is anticipated that the first amendment stage of BS 7671:2008 will include Section 444, *Measures against electromagnetic influences*, which is currently in development.

3.9.1 Protection of low voltage installations against temporary overvoltages due to earth faults in the high voltage system and due to faults in the low voltage system

Generally, Section 442 has requirements for the designers of substations where, depending on the type of earthing systems, faults on the HV side of a distribution transformer can be projected across to the LV side, or vice versa.

Where the substation is installed, owned and operated by the distributor then Section 442 is of no consequence to the designer of the low voltage installation. Should a privately owned substation be installed then IEC 61936-1 *Power installations exceeding 1 kV a.c. – Part 1: Common rules* should be consulted.

The following are four situations which generally cause the most severe temporary overvoltages:

442.2 **1** Fault between the high voltage system(s) and Earth
442.3 **2** Loss of the neutral in a low voltage system
442.4 **3** Accidental earthing of a low voltage IT system
442.5 **4** Short-circuit in the low voltage installation.

3.9.2 Protection against overvoltages of atmospheric origin or due to switching

Section 443

This section of BS 7671 deals in part with transient overvoltages of atmospheric origin (i.e. lightning strikes). The condition of external influences which exists in the United Kingdom is AQ1 (see below for an explanation), which means that the UK is in the fortunate position of not having to install surge protective devices to protect the wiring and equipment comprising a fixed installation, under normal circumstances.

131.6.2 This is a specific application of the fundamental requirement of Regulation 131.6.2 for protection against overvoltage. In addition, Regulation 133.4 requires electrical
133.4 equipment to be selected so that it will not cause harmful effects on other equipment or impair the supply during normal service, including switching operations.

BS EN 62305 Lightning protection of buildings, which is excluded from the scope of BS 7671, is covered by BS EN 62305 *Protection against lightning.*

3.9.3 AQ criteria

Appx 5 Overvoltages of atmospheric origin are categorised in Appendix 5 of BS 7671 as:

AQ1 – Negligible (less than or equal to 25 thunderstorm days per year)
AQ2 – Indirect exposure (more than 25 thunderstorm days per year)
AQ3 – Direct exposure.

A risk assessment method may be used instead of the AQ criteria in determining whether surge protection should be used. The following are different consequential levels of protection: 443.2.4

1. Consequences related to human life, e.g. safety services, medical equipment in hospitals
2. Consequences related to public services, e.g. loss of public services, IT centres, museums
3. Consequences to commercial or industry activity, e.g. hotels, banks, industries, commercial markets, farms
4. Consequences to groups of individuals, e.g. large residential buildings, churches, offices, schools
5. Consequences to individuals, e.g. residential buildings, small offices.

For levels of consequences 1 to 3, protection against overvoltage must be provided.

There is no need to perform a risk assessment calculation for levels of consequences 1 to 3 because this calculation always leads to the result that the protection is required.

3.9.4 Equipment categories

To comply with the requirements of Chapter 44, relevant equipment product standards need to have at least the values of withstand voltage of Table 44.3. Examples of the type of equipment in each category are given in Table 44.4. Table 44.3 Table 44.4

Category IV equipment (equipment with very high impulse voltage) has a required minimum impulse withstand voltage of 6 kV, for a nominal voltage of 230 V. As explained in Table 44.4, such equipment is to be used at or in the proximity of the origin of the electrical installation, upstream of the main distribution board. As such equipment is installed at or near the origin of the supply, e.g. an electricity meter, primary overcurrent device or ripple control unit, any transient overvoltage is likely to be at a high value due to there being little or no attenuation.

Category I equipment (equipment with reduced impulse voltage) has a required minimum impulse withstand voltage of 1.5 kV, for a nominal voltage of 230 V, and is not to be connected to the electricity supply without surge protection. Table 44.4 comments that Category I equipment is intended to be connected to the fixed electrical installation where protection against transient overvoltage is external to the equipment.

Where protection is not required

In the following situations, overvoltage protective devices are not required for protection of an installation against overvoltage of atmospheric origin. The only requirement is that the impulse withstand voltage of equipment is in accordance with Table 44.3: Table 44.3

- where an installation is supplied by a low voltage system containing no overhead lines (a suspended cable having insulated conductors with earthed metallic covering or insulated conductors twisted together is deemed to be an underground cable for the purposes of Chapter 44)
- where the condition of external influences AQ1 exists and an installation is supplied by a low voltage network which includes overhead lines, or where an installation includes an overhead line.

As mentioned in section 3.9.2, the condition of external influences AQ1 exists in the United Kingdom.

Where protection is required

Protection of an installation against voltages of atmospheric origin must be provided if the ceraunic level at a location corresponds to the condition of external influences AQ2 and the installation is supplied by, or includes, a low voltage overhead line.

A lightning flash density greater than 2.24 flashes/km^2/year corresponds to condition AQ2 and although the mean flash density in the UK does not exceed 1.0 flash/km^2/year, the flash may be concentrated at a particular location which, as a consequence, will be subject to a higher than expected risk of experiencing a high level of overvoltage. Installations associated with tall isolated structures, for example radio transmitter towers, may require protection if situated in an area where the lightning flash density is 1.0 flash/km^2/year. A risk assessment of structures in areas of high lightning activity in the UK is necessary in order to determine whether or not protection of an installation is required.

Where protection against overvoltages of atmospheric origin is required, such protection must be provided in the building by:

1 a surge protective device with a protection level not exceeding Category II, or
2 other means providing at least an equivalent attenuation of overvoltages.

443.2.6 Overvoltage protective devices should be located as close as possible to the origin of the installation.

3.9.5 Surge protective devices (SPDs)

A surge protective device (SPD) is a device that is intended to limit transient overvoltages and divert surge currents. SPDs must have the necessary capability to deal with the current levels and durations involved in the switching surges to be expected at their point of installation.

In most cases, switching overvoltages are less damaging than lightning overvoltages, and SPDs that are effective for protection against lightning overvoltages are also effective against switching surges.

3.9.6 Switching overvoltages

Generally, any switching operation, fault initiation, interruption, etc., in an electrical installation is followed by a transient phenomenon in which overvoltages can occur. The sudden change in the system can initiate damped oscillations with high frequencies (determined by the resonant frequencies of the network), until the system is stabilised to its new steady state. The magnitude of the switching overvoltages depends on several parameters, such as the type of circuit, the kind of switching operation (closing, opening, restriking), the nature of the loads and the protective device. In most cases, the maximum overvoltage is up to twice the amplitude of the system voltage but higher values can occur, especially when switching inductive loads (motors, transformers) or capacitive loads, or even resistive loads connected very near to the terminals of a supply transformer. Also, interruption of short-circuit currents can cause high overvoltages. If current chopping occurs, relatively high energy can be stored in inductive loads and oscillations can occur on the load side of the opening switch or protective device.

External influences 4

4.1 External influences

The effect of environmental conditions and general characteristics around various parts of the installation must be assessed to enable suitable electrical equipment to be specified. 301.1

A list of external influences relating to Chapter 32 of BS 7671 is given in Chapter 51 and Appendix 5. Many of these external influences are also identified in Section 522. Chapter 52 of BS 7671 applies to the selection and erection of wiring systems but the classification of external influences is applicable and relevant to all types of electrical equipment. Chap 51 Appx 5 Section 522

If every part of the wiring system complies with the IP degree of protection for the worst circumstances expected in the particular location and correct measures are taken during installation and maintenance, the selection and erection requirements of BS 7671 with respect to external influences will generally be satisfied. Due consideration must be given to the installation of equipment, including fixings, and manufacturers' installation instructions complied with, or the declared IP rating of the equipment may not be achieved. Good workmanship is, as always, required. 134.1.1

Equipment must be of a design appropriate to the situation in which it is to be used or its mode of installation shall take account of the conditions likely to be encountered. 512.2.1

All electrical equipment selected must be suitable for its location of use and method of installation and equipment should not be modified on site unless it is designed and manufactured to allow this. Manufacturers' recommendations and instructions should be followed. BS 7540-2:2005 *Guide to use of cables with a rated voltage not exceeding 450/750 V* gives general cable selection guidance.

Equipment must also be protected against damage and ingress of foreign bodies during construction work. This damage and ingress may be more severe than operational conditions.

References are made in the following sections to certain aspects covered in Section 522. The coding used in Appendix 5 of BS 7671 is given in parentheses.

4.2 Ambient temperature (AA)

Local ambient temperature means the temperature of the air or other medium around the equipment. Part 2

There can be a wide range between the highest and lowest temperatures around equipment, depending on time of year, artificial heating, etc. A design must be based 522.1

on the most extreme ambient envisaged. In the UK, external ambient temperatures between -10 °C and +35 °C commonly occur and occasionally even lower or higher temperatures than these.

Appx 4 Table 4B1 Table 4B2 The cable current-carrying capacity tables in Appendix 4 of BS 7671 are based on an ambient air temperature of 30 °C and, for buried cables, an ambient ground temperature of 20 °C. Where these temperatures differ, rating factors are to be used. Appendix 4, clause 2.1 and Tables 4B1 and 4B2 should be consulted.

522.1 Although the temperature in most installations in the UK would not usually be over 30°C, close to a domestic heating appliance or in an industrial environment it may be higher. BS 1363 accessories and cable with an operating temperature rating of 70 °C or lower are not recommended for installation in areas of higher ambient temperature such as airing cupboards and in particular must not be covered with clothes or other items. Heat-resisting cables appropriate to the ambient and equipment temperatures must be utilized.

Cables in heated floors or other parts must be suitable for the highest temperature reached.

Cables and equipment must also be suitable for the lowest likely temperature. In particular, general-purpose thermoplastic cables should not be installed in cold places. Attention is drawn to the fact that, as the temperature decreases, thermoplastic compounds become increasingly stiff and brittle, with the result that, if the cable is bent quickly to a small radius or is struck at temperatures in the region of 0 °C or lower, there is a risk of shattering the thermoplastic components.

To avoid the risk of damage during handling, it is desirable that thermoplastic insulated or sheathed cables should be installed only when both the cable and the ambient temperature are above 0 °C and have been so for the previous 24 hours, or where special precautions have been taken to maintain the cables above this temperature. BS 7540 advises a minimum ambient temperature of 5 °C for some types of thermoplastic insulated and sheathed cables.

4.3 Solar radiation (AN) and ultraviolet radiation

522.2.1 522.11.1 Where cables are installed in situations subject to solar gain or other radiant heat sources, they should be resistant to damage from that source or effectively protected from damage. Cables may need to be derated where there are heat sources or heat is absorbed by solar gain.

To protect cables from the adverse effects of ultraviolet (UV) radiation, the sheaths often contain carbon black or alternative uv stabilisers. Physically shielding the cables from direct solar radiation may also be necessary. In situations where high ambient temperatures are present (e.g. hot countries) cables will need to be derated. Shielding may be necessary, to protect from both direct heat and heat absorbed from solar radiation but cables should be freely ventilated. Cable manufacturers will provide guidance on the use of cables outdoors.

Where weather resistance is required, the type of cable sheath to give protection should be carefully considered. Cables sheathed with thermoplastic to BS 6746 or synthetic thermosetting to BS 6899 can give adequate performance but those sheathed with natural thermosetting do not perform well. Cable sheathing materials should conform to BS 7655. Protection from mechanical damage may also be required. Further advice is given in BS 7540.

The exposure of thermoplastic insulation or sheathing to high temperatures for any length of time will lead to softening. If such a temperature is maintained for long periods, the thermoplastic can decompose and give off corrosive products which may attack the conductors and other metalwork. Softening of the thermoplastic may also allow the conductor to move through the insulation if any mechanical force is applied (e.g. cables hanging vertically). Decentralisation of conductors may also take place where cables are overheated when bunched inside trunking.

4.4 The IP and IK classifications

The designer must assess each area of external influence then ensure the required classification is selected and applied correctly. 512.2

The IP classification code, BS EN 60529:2004, describes a system for classifying the degrees of protection provided by the enclosures of electrical equipment. Superseded descriptions such as 'ordinary', 'drip-proof', 'splash-proof' and 'water-tight', are not defined terms.

The IK classification standard BS EN 62262 describes a system for classifying the degrees of protection provided by enclosures for electrical equipment against external mechanical impacts. The letters IK are followed by two numerals which identify a specific impact energy.

Appendix B of this Guidance Note lists the degrees of protection indicated by these classifications.

4.5 Presence of water (AD) or high humidity (AB)

Any wiring system or equipment selected and installed must be suitable for its location and able to operate satisfactorily without deterioration during its working life. The presence of water can occur in several ways, e.g. rain, splashing, steam/humidity, condensation, and at each location where it is expected to be present its effects must be considered. Suitable protection must be provided, both during construction and for the completed installation. 522.3

In damp situations and wherever they are exposed to the weather, all metal sheaths and armour of cables, metal conduit, ducts, trunking, clips and their fixings should be of corrosion-resistant material or finish. They should not be in contact with other metals with which they are liable to set up electrolytic action (see section 4.7 also). 522.5

In any situation and during installation, the exposed insulation at terminations and joints of cables which are insulated with hygroscopic material, e.g. mineral insulated cable, must be sealed against the ingress of moisture. Such sealing material and any material used to insulate the conductors where they emerge from the cable insulation should have adequate insulating and moisture-proofing properties and retain those properties throughout the range of service temperatures.

The manufacturer's recommendations regarding the termination and installation of all types of cable must be strictly observed.

Section 526 In a damp situation, enclosures for cores of sheathed cables from which the sheath has been removed and for non-sheathed cables at terminations of conduit, duct, ducting or trunking systems should be damp-proof and corrosion-resistant. Every joint in a cable should be suitably protected against the effects of moisture.

522.3.2 Conduit systems not designed to be sealed should be provided with drainage outlets at any points in the installation where water might otherwise collect. Those outlets must not affect the electrical safety of the system, or allow vermin access.

512.2.1 If it is necessary to locate a switch, socket-outlet, etc. near to a sink, the device must be protected from any water splashing that may occur. This can be achieved either by locating the accessory in a position where it will not be exposed to splashing, or providing a splash-resistant covering, or providing a weather protected device.

(Further information is contained in Appendix C.6.2, 'Sinks and electrical accessories'.)

It may occasionally be necessary to consider the installation of cables in locations that permanently contain water, such as fountains and marinas. It must be noted that the plastic insulating and sheathing materials are not totally waterproof and will absorb some water over time. Such cables are not suitable for continuous immersion and the cable manufacturer's advice should be taken for such installations. (See Guidance Note 7 for further information.)

4.6 Presence of solid foreign bodies (AE)

522.4 Where dusty conditions may be expected to occur, equipment and enclosures for conductors and their joints and terminations should have the degree of protection of at least IP5X. There are two conditions recognised in this particular IP Code, the first condition being where the normal working cycle of the equipment concerned causes reductions in air pressure, e.g. thermal cycling effects, and the second condition being where such pressure reductions do not occur. These conditions should be considered during the design and selection procedure and appropriate equipment chosen.

It is necessary to have good operational housekeeping policies to ensure dust is kept to a minimum. A build-up of dust on electrical equipment, cables, etc., can act as thermal insulation and cause overheating of the equipment or cable. The dust may even be of a type that can ignite if the temperature of equipment rises. However, combustible dusts are outside the scope of this Guidance Note and reference should be made to BS EN 50281 and BS EN 61241 for further guidance.

529.3 To assist in cleaning in dusty installations, a design should be developed in which inaccessible surfaces and locations are minimized and surfaces that can collect dust are made as small as practicable, or sloped to shed dust. Cable ladder rack is preferred to cable tray and ladder rack should be suspended vertically (i.e. on edge) rather than horizontally (i.e. flat).

4.7 Presence of corrosive or polluting substances (AF)

522.5 In damp situations, where metal cable sheaths and armour of cables, metal conduit and conduit fittings, metal ducting and trunking systems and associated metal fixings are liable to chemical or electrolytic attack by materials of a structure with which they may come into contact, it is necessary to take suitable precautions against corrosion, such as galvanizing or plating.

Materials likely to cause such attack include:

- materials containing magnesium chloride which are used in the construction of floors and plaster mouldings
- plaster coats containing corrosive salts
- lime, cement and plaster, for example on unpainted walls, or over cables buried in chases
- oak and other acidic woods
- dissimilar metals liable to set up electrolytic action.

Application of suitable coatings before erection, or prevention of contact by separation with plastics, are recognised as effective precautions against corrosion.

Contact between bare aluminium conductors, aluminium sheaths or aluminium conduits and any parts made of brass or other metal having a high copper content should be avoided, especially in damp situations, unless the parts are suitably plated. If such contact is unavoidable, the joint should be completely protected against ingress of moisture. Modern, friction welded bi-metallic joints are available for most tape to tape or tape to cable joints. Wiped joints in aluminium sheathed cables should always be protected against moisture by a suitable paint, by an impervious tape or by embedding in bitumen.

Bare copper sheathed cables should not be laid in contact with zinc plated (galvanized) materials such as cable tray in damp conditions. This is because the electropotential series indicates that zinc is anodic to copper and therefore preferential corrosion of the zinc plating may occur. This action will not affect the copper but may cause corrosion of the cable tray. The presence of moisture is essential to produce electrolytic action, so in dry conditions this action will not occur. If moisture is present then electrolytic action will take place but the extent of any corrosion is dependent upon the relative areas of the two metals and the conductivity of the electrolyte (moisture in this instance). In these circumstances, thermoplastic sheathing should be used.

Hostile environments and chemicals also can attack the conductor, its insulation and sheath and any enclosure or equipment.

A few examples of this are:

- petroleum products, creosote, some solvents and hydrocarbons attack rubber and may attack polymeric materials such as thermoplastic
- plasticisers migrate to polystyrene from thermoplastic and also to some types of plaster (see section 7.5 also)
- hostile atmospheres (e.g. those in the vicinity of plastics processing machinery and vulcanising which produces a sulphurous atmosphere)
- coastal areas with salt-laden air
- in areas where animals are kept (agricultural environments, kennels, etc.) animal urine can be corrosive.

In such cases the manufacturer's advice should be obtained and care taken in the measures employed.

To provide adequate protection from corrosion, the corrosive substances need to be identified and the manufacturer's or a specialist's advice obtained.

4.8 Impact (AG), vibration (AH) and other mechanical stresses (AJ)

522.6 Any part of the fixed installation which may be exposed to mechanical damage must be able to survive it. Typical examples of mechanical damage will include impacts from external objects, abrasion of insulation, penetration, etc. In workshops where heavy objects are moved, the traffic routes should be avoided. If it is not possible to avoid the traffic routes, heavy duty equipment or localised protection must be provided.

See Appendix B for details of the IK code for classification of the degrees of protection by enclosures against impact.

522.7 Final connections to plant that is adjustable or produces vibration must be designed to accommodate this and a final connection made in flexible cables or flexible conduit or a properly supported cable vibration loop allowed.

522.8 Allowance must be made for the thermal expansion and contraction of long runs of steel or plastic conduit or trunking and adequate cable slack provided to allow free movement. The expansion or contraction of plastic conduit or trunking is somewhat greater than that of steel for the same temperature change.

For busbar trunking systems, BS EN 60439-2 identifies a busbar trunking unit for building movements. Such a unit allows for the movement of a building due to thermal expansion and contraction. Reference should be made to the manufacturer since requirements may differ with respect to design, current rating or orientation of the busbar trunking system, i.e. riser or horizontal distribution.

522.15 Cables crossing a building expansion joint should be installed with adequate slack to allow movement and a gap left in any supporting tray or steelwork. A flexible joint should be provided in conduit or trunking systems.

4.9 Presence of fauna (AL), flora and/or mould growth (AK)

Section 705 Special requirements for animal housing areas will depend on the type and size of animal and its required environment. Section 705 gives some guidance on farm requirements but specialist advice will be needed for more unusual species.

522.9
522.10 Cables and equipment may be subject to attack and damage from plants and animals as well as the environment. The damage may be caused by such diverse occurrences as vermin chewing cables, insect or vermin entry into equipment, physical impact/damage by larger animals such as can occur in agricultural areas and plant growth placing excessive strains on equipment over a period of time – a small lighting column base can be moved by tree roots – or choking equipment and blocking ventilation. Rodents have a particular taste for some forms of cable sheath and can gnaw through the cable sheath and insulation to expose conductors. They build nests and the nests are usually constructed of flammable material. Such a combination is ideal for the propagation of fire where the nest surrounds the wiring material. Cables impregnated with anti-vermin compounds have not been found to be successful and may not comply with Health and Safety legislation. Wherever possible, cables should not be routed on likely vermin 'runs', e.g. on the tops of walls or in voids, etc., but located in full view for easy inspection for damage.

As far as practicable, cables and equipment should be installed away from areas or routes used by animals or be of a type to withstand such attack. Thermoplastic sheathing and insulation may be chewed by vermin and in such areas steel conduit may be required.

The access of insects is difficult to prevent as they can enter through small gaps such as vent holes. Equipment and wiring systems in such locations must be carefully sealed and any vents fitted with breathers, etc.

In specific cases, such as installations in properties with thatched roofs, damage from animals is more likely than in conventional buildings; the use of steel conduit/trunking and mineral insulated sheathed cable (MICC) in such installations is advisable.

For more information on installations in buildings with thatched roofs, see Guidance Note 4 and refer to 'The Dorset Model' issued by the Dorset Building Control Technical Committee. The Dorset Model offers a uniform approach to thatched buildings, which is being advocated across Dorset where compensatory requirements are considered acceptable to achieve compliance with the Building Regulations.

4.10 Potentially explosive atmospheres

The selection and erection of equipment for installations in areas with potentially explosive atmospheres of explosive gases and vapours or combustible dusts is outside the scope of this Guidance Note. The design and the selection and erection of equipment for such installations requires specialist knowledge.

Further information may be found in BS EN 60079 *Electrical apparatus for explosive gas atmospheres* and BS EN 61241-0:2006 *Electrical apparatus for use in the presence of combustible dust, General requirements*. The Dangerous Substances and Explosive Atmospheres Regulations 2002 give the statutory requirements. Appx 2, 5(ii)

4.11 Precautions where particular risks of fire exist

Section 422 of BS 7671 specifies particular requirements for electrical installations in: Section 422 422.1

1 emergency escape routes in buildings 422.2
2 locations with risks of fire due to the nature of processed or stored materials 422.3
3 buildings constructed mainly of combustible materials (e.g. wood) 422.4
4 fire propagating structures (e.g. high-rise buildings) 422.5
5 locations of national, commercial, industrial or public significance. 422.6

Specialist advice may need to be sought at the design stage.

Regarding item 1 above, detailed guidance on the requirements for escape routes is given in Chapter 4 of Guidance Note 4: *Protection Against Fire*. GN4

In the above locations, excepting certain wiring systems, electrical equipment must be restricted to that necessary to the use of the location and be suitably selected and erected so as not to be itself a cause of fire in normal service or during a fault. 422.3.5 422.1.1 422.1.2

4.11.1 Locations with risks of fire due to the nature of processed or stored materials

A clear distinction has to be made between installations in locations with explosion risks and locations with risk of fire due to the nature of the process undertaken in the location or because of the nature of stored materials.

If the risk is from the process or stored materials and there is no explosion risk, the following must be followed:

422.3 **1** restrict electrical equipment to only that necessary for the location

422.3.8 422.3.4 **2** select only equipment appropriate for the location, e.g. of suitable IP rating

3 where wiring is not completely embedded in non-combustible material, such as plaster or concrete, select wiring systems with non-fire propagation characteristics, e.g. cables enclosed in conduit complying with the non-flame propagating requirements of BS EN 61386-21 or ducting or trunking complying with BS EN 50085 or mineral insulated sheathed cable (MICC). Cable types complying with BS EN 60332-1-2 are acceptable

422.3.5 **4** if, of necessity, a wiring system traverses such a location but is not necessary to the use of the location, then the cabling should have no connections unless in an enclosure meeting the fire test requirements of BS 6458-2.1 (IEC 60695-2-1:1980) with the following test temperatures:

BS 6458
- 750 °C for surface mounted installations,
- 850 °C for installation in combustible hollow walls

433.2.1 433.3.1 **5** place overload and short-circuit protective devices upstream and outside the location

422.3.9 **6** except for mineral insulated cables and busbar trunking or powertrack systems, fit an RCD with a rated residual operating current of 300 mA or less. Where there is concern that quite low current faults could cause fire, the rated residual operating current should not exceed 30 mA

422.3.12 **7** provide all wiring systems with a protective conductor. PEN conductors must not be installed for equipment within the location

8 install an isolating device (disconnector) outside the location to switch off all supplies to the location

422.3.13 **9** always switch the neutral as well as the line conductors when isolating equipment

422.3.5 **10** don't install bare live conductors for sliding contact wiring except where there is no practical alternative, e.g. sliding contacts for a hoist when either an enclosed system or flexible cables should be used

422.3.14 **11** use only rubber sheathed cables to BS 6008 type HO7 RN-F or equivalent for flexible wiring

BS 6008 **12** if SELV or PELV circuits are installed, contain all live parts within enclosures to IP5X or provide insulation

13 install all switchgear outside the location unless it is completely enclosed by non-combustible material

422.3.7 **14** protect all motors (that are not continuously supervised) against excessive temperature by overload protective devices with manual resetting or equivalent

15 protect motors with: 422.3.7

- star-delta starting without automatic changeover from star to delta against excessive temperature in the star mode in addition to delta
- slipring starters from being left with resistance in the rotor circuit

16 use luminaires with limited surface temperatures only. Fittings manufactured to BS EN 60598 have surface temperatures limited to 90 °C under normal conditions and 115 °C under fault 422.3.2 422.3.8 BS EN 60598

17 avoid the use of small spotlights and projectors. Any that are necessary should be separated from combustible materials by minimum distances as follows: 422.3.1

- up to 100 W – 0.5 m
- over 100 W up to 300 W – 0.8 m
- over 300 W up to 500 W – 1 m

Note: Greater distances may be necessary if the lamps are focused on combustible material.

18 take precautions against components that are likely to run hot, such as lamps, from falling out of luminaires 422.3.8

19 fix all heating appliances, and for storage types select only ones which will prevent the ignition of combustible dust or fibres 422.3.15 422.3.16

20 ensure by selection that, in normal operation, enclosures of thermal appliances such as heaters do not attain a temperature higher than 90 °C. 422.3.2

4.11.2 Locations constructed of combustible materials

Electrical equipment which is to be installed on or in a combustible wall must: 422.4

1 comply with the relevant standard. There is a requirement on manufacturers to ensure that electrical equipment meets the fire resistance tests within the standards which cover their products; or 422.4.3

2 be enclosed with a suitable thickness of non-flammable material. As such materials generally provide excellent heat insulation, this could have the effect of making the interior of the enclosure hotter where flush-mounted equipment is separated from the building material by additional thermal insulation. 422.4.4

Cables and cords

Cables and cords must comply with the flame tests in BS EN 60332-1-2:2004 *Procedures for 1 kW pre-mixed flame*, which are carried out by manufacturers. 422.4.5

Cables manufactured to the following standards all meet the flame tests:

BS 5467, BS 6004, BS 6346, BS 6724, BS 6883, BS 7211, BS 7629, BS 7846, BS 7889 and BS 7919.

(See Table F.1 in Appendix F for a comparison of harmonized cable types to BS 6004.)

It would be prudent to check with the wholesaler to ensure, before purchasing cables, that they comply with one of these Standards.

Conduit and trunking systems

422.4.6 Conduit systems must be in accordance with BS EN 61386-1:2004 *General requirements*. Trunking systems and ducting systems must be in accordance with BS EN 50085-1:1999 *General requirements*. There is a requirement on manufacturers to ensure that conduit, trunking and ducting systems meet the fire resistance tests within these standards.

Further information on both standards is contained in Section 5.1 of this Guidance Note.

4.11.3 Fire propagating structures

422.5.1 In structures or buildings where the shape and dimensions are such as will facilitate the spread of fire (e.g. chimney effect), precautions are required to be taken to ensure that the electrical installation cannot propagate a fire.

Section 527 Suitable cables and wiring systems must be installed and a high standard of sealing around and, as appropriate, within wiring systems are essential to minimise the risk of propagation of fire.

4.11.4 Locations of national, commercial, industrial or public significance

422.6 Regulation 422.6 states that the requirements of Regulation 422.1 shall apply to
422.1 locations that include buildings or rooms with assets of significant value and gives examples that include national monuments and museums. Other examples given of buildings having commercial, industrial or public significance are railway stations and airports, laboratories, computer centres and certain industrial and storage facilities.

In these locations protective measures could include the installation of mineral insulated cables to BS EN 60702, or cables with improved fire-resisting characteristics (see Chapter 5).

Improved fire protection may be required by the use of reactive fire protection systems, e.g. sprinkler systems.

Installation of cables 5

5.1 Cable selection

Cables must be selected to comply with the electrical characteristics of current rating, voltage drop, etc., as required by Parts 4 and 5 of BS 7671 and the physical protection characteristics as required by Chapter 52. The correct selection and erection of a complete wiring system is necessary to provide compliance with Chapter 52.

Over-specification of cable performance requirements can result in increased installation costs and the designer must assess performance reasonably.

Cables must be selected to perform their required function in the environment throughout their expected life but there is no justification for overdesign. Consideration of the overall lifetime costs of a cable is not overdesign. There may be a trade-off to increase the size of a cable to reduce overall energy losses in the cable during its life. Regular periodic inspection (and testing as necessary) of an installation must be carried out to monitor cable and equipment conditions and this should identify changes of use of an installation that would require modifications to the installation during its life. (See Guidance Note 3 for further information.)

BS 7450:2005 (series) *Method for determination of economic operation of power cable size based on overall lifetime costs (including cable and installation costs and the cost of waste energy)* may also be of assistance in the determination of optimum conductor sizes based on energy losses.

Cables in emergency and life safety systems must be able to perform their intended function in emergency situations such as fire. The requirements that cables will have to meet for circuit integrity to be maintained under fire conditions are stated in the relevant system specification:

- BS 5266-1:2005 *Emergency lighting* – Part 1: Code of practice for the emergency lighting of premises.
- BS 5839-1:2002 +A2:2008 *Fire detection and fire alarm systems for buildings* – Part 1: Code of practice for system design, installation, commissioning and maintenance.
- BS 7346-6:2005 *Components for smoke and heat control systems* – Part 6: Specification for cable systems.
- Guidance Note 4: *Protection Against Fire* for further information on cable selection for fire protection and alarm systems.
- BS 7540-1:2005 *Guide to use of cables with a rated voltage not exceeding 450/750 V* gives guidance on the general use of some types of cable, mainly the general wiring types (see also Appendix F).

527.1.3 527.1.4 General building wiring cables that comply with the requirements of BS EN 60332-1-2:2004 (formerly BS 4066-1:1995) for flame propagation are considered to have a suitable non-flame propagating property to allow their installation in buildings and between fire zones or compartments without further physical fire protection, unless there is significant risk of fire e.g. hazardous or flammable materials stored or special processes. Cables not to BS EN 60332-1-2 may not pass unprotected between building fire compartments or zones and should generally only be used in short lengths and not in areas of fire risk.

422.3.4 Most types of general cable comply with BS EN 60332-1-2, including thermoplastic insulated and sheathed cables to BS 6004 and thermosetting/SWA/thermoplastic cables to BS 5467. A more severe test is prescribed in BS 50266 series of specifications (formerly BS 4066-3), which tests flame propagation of a group of cables and it may be that cables installed in groups in risers, or control and communications cables where bunched, should be required to comply with this standard.

5.1.1 Conduit, trunking and ducting systems standards

The BS EN 61386 series of standards covering conduit systems for electrical installations has replaced the following earlier standards:

- BS 731-1:1952 (1993) *Flexible steel conduits*
- BS 4568 *Steel conduit and fittings with metric threads*
- BS 4607 *Non-metallic conduit and fittings*
- BS 6099 *Conduits for electrical installations*
- BS EN 50086-1:1994 *Specification for conduit systems for cable management. General requirements*
- BS EN 50086-2-1:1996 *Specification for conduit systems for cable management. Particular requirements. Rigid conduit systems*
- BS EN 50086-2-2:1996 *Specification for conduit systems for cable management. Particular requirements. Pliable conduit systems*
- BS EN 50086-2-3:1996 *Specification for conduit systems for cable management. Particular requirements. Flexible conduit systems.*

BS EN 61386 series of standards is in various parts; the following parts have been published:

- BS EN 61386-1:2008 *Conduit systems for cable management. General requirements*
- BS EN 61386-1:2004 *Conduit systems for cable management. General requirements* (remains current until June 2011 when it is replaced by the 2008 version) – see Table 5.1 of this Guidance Note
- BS EN 61386-21:2004 *Conduit systems for cable management. Particular requirements. Rigid conduit systems*
- BS EN 61386-22:2004 *Conduit systems for cable management. Particular requirements. Pliable conduit systems*
- BS EN 61386-23:2004 *Conduit systems for cable management. Particular requirements. Flexible conduit systems.*

Conduit systems buried underground are still covered by:

- BS EN 50086-2-4:1994 *Particular requirements for conduit systems buried underground.*

The latest standards should always be used whenever possible.

The designer must select an appropriate level of performance for application for the specific characteristics of the installation and, again, over-specification in design will increase costs. A listing of performance criteria for conduit is given in Table 5.1.

▼ **Table 5.1** Performance requirements in BS EN 61386-1 *Conduit systems for cable management. General requirements*

Performance characteristics	Performance classification							
	0	1	2	3	4	5	6	7
Compression (N)	0	125	320	750	1250	4000	–	–
Impact energy (joules)	–	4.9	9.8	20	59	196	–	–
Low temperature (°C)	–	+5	-5	-15	-25	-45	–	–
High temperature (°C)	–	+60	+90	+105	+120	+150	+250	+400
IP rating								
Solid object	–	–	–	2.5 mm	1.0 mm	dust protected	dust tight	–
Water	0	drops	drops at 15°	spray	splashing	jets	powerful jets	temporary immersion
Corrosion protection	–	low inside and outside	medium inside and outside	medium inside, high outside	high inside and outside	–	–	–
Tensile strength (N)	0	100	250	500	1000	2500	–	–
Suspended load (N)	0	20	30	150	450	850	–	–

Specific flame propagation tests are included in the conduit standards. Flame propagation is defined in terms of a pass or fail test based on temperatures consistent with human safety (temperatures above which persons cannot survive). Materials that pass the test are identified as 'non flame propagating' and are required to be marked with a clearly legible identification or may be of any colour except yellow, orange or red, unless clearly marked on the product to be of non-flame propagating material. Flame propagating conduit must be coloured orange. 527.1.6

The BS EN 50085 series of standards, which covers cable trunking systems and cable ducting systems, is in various parts; the following parts have been published:

- BS EN 50085-1:2005 *Cable trunking systems and cable ducting systems for electrical installations. General requirements*
- BS EN 50085-1:1999 *Cable trunking systems and cable ducting systems for electrical installations. General requirements* (remains current for Part 2–3)
- BS EN 50085-2-1:2006 *Cable trunking systems and cable ducting systems for electrical installations. Cable trunking systems and cable ducting systems intended for mounting on walls and ceilings*
- BS EN 50085-2-2:2009 *Cable trunking systems and cable ducting systems for electrical installations – Part 2-2: Particular requirements for cable trunking systems and cable ducting systems intended for mounting underfloor, flush floor or on floor* (to be published in 2009)
- BS EN 50085-2-3:2001 *Cable trunking and cable ducting systems for electrical installations. Particular requirements for slotted cable trunking systems intended for installation in cabinets. Section 3: Slotted in cabinets.*

The following standards are still current but will eventually be replaced by the BS EN 50085 standards:

- BS 4678-2:1973 *Cable Trunking. Steel underfloor (duct) trunking*
- BS 4678-4:1982 *Cable trunking. Specification for cable trunking made of insulating material.*

All the latest Harmonized Standards are performance related and allow materials that will satisfy the required performance classification and, while these new standards do not change any existing material specifications, they provide alignment of requirements throughout Europe. It is no longer correct to mention 'material specification' in relation to these standards. Products meeting the Harmonized Standards may then carry the 'CE' mark. However, it will be necessary to check the technical literature or with the manufacturer for special applications.

5.2 Cables concealed in structures

5.2.1 Walls and partitions

522.6.6 A cable concealed in a wall or partition at a depth of less than 50 mm from a surface of the wall or partition must:

1 incorporate an earthed metallic covering which complies with the requirements of BS 7671 for a protective conductor of the circuit concerned, the cable complying with BS 5467, BS 6346, BS 6724, BS 7846, BS EN 60702-1 or BS 8436, or
2 be enclosed in earthed conduit complying with BS EN 61386 and satisfying the requirements of BS 7671 for a protective conductor, or
3 be enclosed in earthed trunking or ducting complying with BS EN 50085 and satisfying the requirements of BS 7671 for a protective conductor, or
4 be mechanically protected against damage sufficient to prevent penetration of the cable by nails, screws and the like, or
5 be installed in a zone within 150 mm from the top of the wall or partition or within 150 mm of an angle formed by two adjoining walls or partitions. Where the cable is connected to a point, accessory or switchgear on any surface of the wall or partition, the cable may be installed in a zone either horizontally or vertically, to the point, accessory or switchgear. Where the location of the accessory, point or switchgear can be determined from the reverse side, a zone formed on one side of a wall of 100 mm thickness or less or partition of 100 mm thickness or less extends to the reverse side. These are the so-called 'safe zones'. (See Figure 5.1.)

▼ Figure 5.1
Zones prescribed in Regulation 522.6.6(v)

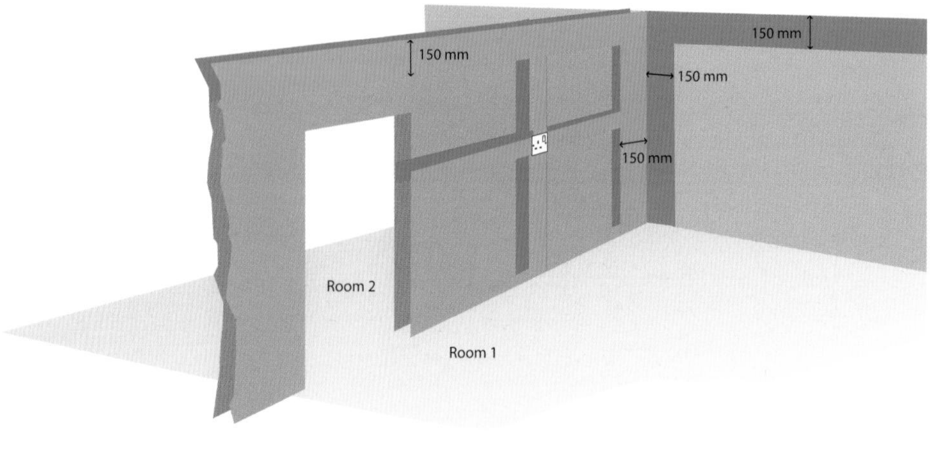

Where a cable is concealed in a wall or partition at a depth of less than 50 mm from a surface of the wall or partition and the installation is not intended to be under the supervision of a skilled or instructed person, a cable installed in the 'safe zones', and not having the mechanical protection in 1, 2, 3 or 4 above, must be provided with additional protection by means of a 30 mA RCD. 522.6.7

Irrespective of the depth of the cable from a surface of the wall or partition, in an installation not intended to be under the supervision of a skilled or instructed person, a cable concealed in a wall or partition the internal construction of which includes metallic parts, other than metallic fixings such as nails, screws and the like, must: 522.6.8

1. incorporate an earthed metallic covering which complies with the requirements of BS 7671 for a protective conductor of the circuit concerned, the cable complying with BS 5467, BS 6346, BS 6724, BS 7846, BS EN 60702-1 or BS 8436, or
2. be enclosed in earthed conduit complying with BS EN 61386 and satisfying the requirements of BS 7671 for a protective conductor, or
3. be enclosed in earthed trunking or ducting complying with BS EN 50085 and satisfying the requirements of BS 7671 for a protective conductor, or
4. be mechanically protected sufficiently to avoid damage to the cable during construction of the wall or partition and during installation of the cable, or
5. be provided with additional protection by means of a 30 mA RCD.

Note: If the cable is installed at a depth of 50 mm or less from the surface of a wall or partition the requirements of Regulation 522.6.6 also apply.

5.2.2 Ceilings and floors

A cable installed under a floor or above a ceiling must be run in such a position that it is not liable to be damaged by contact with the floor or the ceiling or their fixings. A cable passing through a joist within a floor or ceiling construction or through a ceiling support (e.g. under floorboards) must: 522.6.5

1. be at least 50 mm measured vertically from the top, or bottom as appropriate, of the joist or batten, or
2. incorporate an earthed metallic covering which complies with the requirements of BS 7671 for a protective conductor of the circuit concerned, the cable complying with BS 5467, BS 6346, BS 6724, BS 7846, BS EN 60702-1 or BS 8436, or
3. be enclosed in earthed conduit complying with BS EN 61386 and satisfying the requirements of BS 7671 for a protective conductor, or
4. be enclosed in earthed trunking or ducting complying with BS EN 50085 and satisfying the requirements of BS 7671 for a protective conductor, or
5. be mechanically protected against damage sufficient to prevent penetration of the cable by nails, screws and the like.

(See Figures 5.2 and 5.3.)

Notches and drillings in the same joist should be at least 100 mm apart horizontally.

Regulation 522.6.5(v) allows, as an alternative to earthed metal conduit, *'mechanical protection sufficient to prevent penetration of the cable by nails, screws and the like'.* Unless purpose manufactured plates are used, effective protection is very difficult to achieve, bearing in mind modern fixings such as self-tapping screws and shot-fired nails that are available. 522.6.5

▼ **Figure 5.2**
Holes and notches in joists

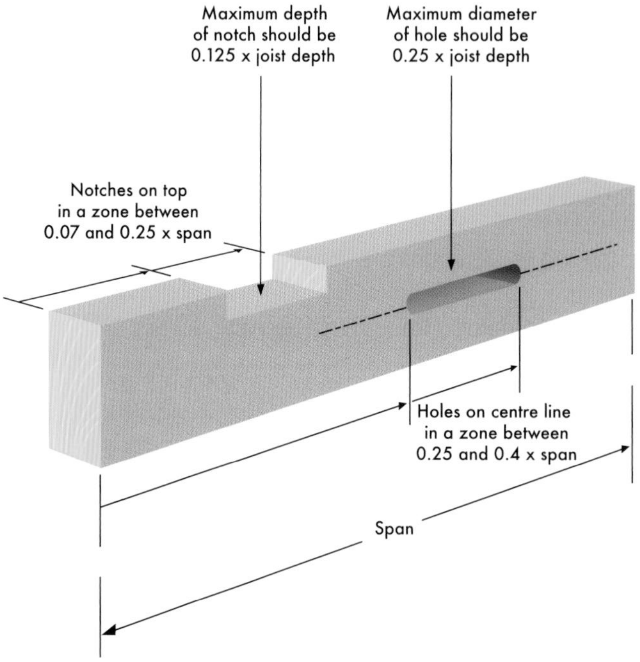

▼ **Figure 5.3**
Cables in floor joists

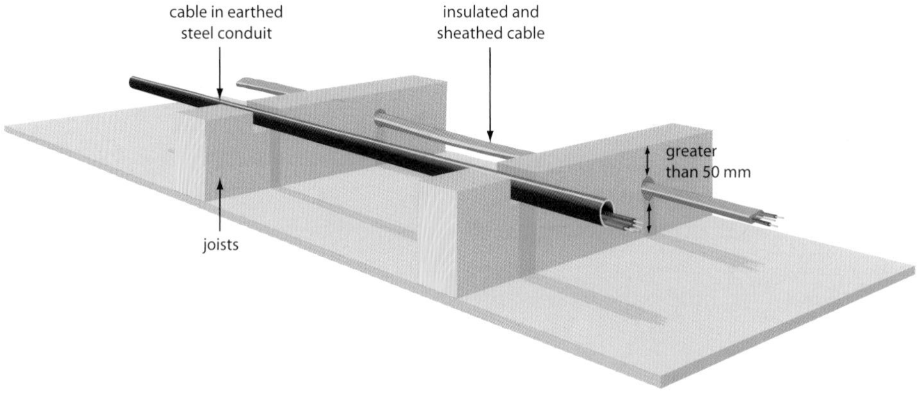

5.3 Cable routes and livestock

Section 705 Particular attention is needed to the routing of cables in locations where livestock is present. (See section 4.9 also.)

Animals in enclosures or buildings may rear up or move as a herd. Such movement can result in severe damage to inappropriately positioned equipment and risks to the animals themselves. Animals may also create corrosive conditions so that normal finishes are not durable enough. Heavy duty equipment with anti-corrosive finishes should be selected and appropriately sited.

Class II installations in heavy duty high impact PVC with suitable Class II accessories and conduit and enclosures have been found to be satisfactory in many cases.

5.4 Capacities of conduit and trunking

522.8 Where cables are installed in conduit or trunking, care is needed to prevent undue strain when installing the cables. A method for dealing with this is given in
Appx A Appendix A. This may be used to determine the minimum size of conduit or trunking necessary to accommodate cables of the same size or mixed sizes.

It should be remembered that cables installed in conduit or trunking must be sized to allow also for the thermal effect of the grouping of such cables. Where advantage is taken of rating factors it may be necessary to indicate this by a local label in the trunking or a note in the installation record documentation to warn future modifiers. Care must be taken when drawing cables into conduit and trunking systems to avoid damage by abrasion.

5.5 Selection of cables and wiring systems with low emission of smoke and corrosive gases when affected by fire

When specifying or selecting cables in areas with high levels of public access, for example hospitals, schools, supermarkets and shopping developments, the designer should consider factors which may affect the safety of the occupants, such as:

1. where the rapid formation of dense, noxious smoke would be likely to produce injury or panic in individuals or crowds
2. along escape routes, where it is important that the rapid but orderly exit of people is not impeded by smoke or fumes
3. where the formation of corrosive halogen-acid products would have a harmful effect upon expensive equipment, i.e. IT equipment.

A list of British Standards is given in Appendix 1 of BS 7671 and includes reference to cables with low emission of smoke and corrosive gases when affected by fire. These include BS 6724, BS 7211, BS 7629, BS 7846 and BS 8436.

Unplasticised thermoplastics, as used in the manufacture of many cable management products (such as conduit, ducting, trunking and cable tray), are non-flame propagating but are not low smoke halogen-free (LSHF) unless they have been specifically ordered as low smoke zero halogen.

It is not expected that general electric cables in the quantities installed in most buildings would provide high levels of smoke on their own in a fire and cables would usually only be expected to burn when other building components are also on fire. However, in some installations, such as public buildings and hospitals, it is preferable to take positive action to limit the production of smoke and fumes. Cables with low emission of smoke and corrosive gases when affected by fire have an important place in the installation design.

Tests for smoke and acid gas emission identify the performance of cables in fire conditions. These tests need to take into account the combination effect of all the materials used in a wiring system and environmental conditions. Specific requirements for smoke and acid gas limits must be specified by the designer and statutory authorities where this is relevant. It must be realised that needless over-specification of cables with low emission of smoke and fumes will impose an unnecessary cost burden on a project.

5.6 'Section 20'

Buildings required to comply with the special provisions of 'Section 20' in London and equivalent provisions in other cities may require special precautions. Cables with a low emission of smoke and corrosive gases when affected by fire, or steel conduit and trunking installations, may be necessary when cables are installed in building voids, such as suspended ceiling voids, where these are used as plenums. Such design requirements should be clarified with the local Fire Officer or Building Control Officer before design commences.

Section 20 of the London Building Acts (Amendment) Act 1939 (as amended primarily by the Building (Inner London) Regulations 1985) is principally concerned with the danger arising from fire within certain classes of building which by reason of height, cubic capacity and/or use necessitate special consideration. The types of building coming within these categories are defined under Section 20 of the amended 1939 Act.

As buildings vary so much in height, cubic capacity, layout, siting, use and construction, the relevant Council will deal with each case on its merits. The principles seek, in certain buildings (or parts of buildings), to ensure the safety of the structure against fire.

Section 20 applies where:

1 a building is to be erected with a storey or part of a storey at a greater height than:
- ▶ 30 m, or
- ▶ 25 m if the area of the building exceeds 930 m^2

2 a building of the warehouse class, or a building or part of a building used for the purposes of trade or manufacture, exceeds 7,100 m^3 in extent unless it is divided by division walls in such a manner that no division of the building is of a volume exceeding 7,100 m^3.

A fire alarm system complying with the current edition of BS 5839-1 should be provided throughout every building except buildings comprising flats and/or maisonettes. These are covered by BS 5839-6 and also may have specific Building Regulations requirements.

In some instances, it may be necessary for such a fire alarm installation to be automatic where the use of the building (or part) warrants it (e.g. in hotels).

In office buildings where the means of escape is based on phased evacuation a number of additional features (such as a public address system) will be necessary.

The fire and rescue service when responding to an emergency in a building should have the indicator panels and associated manual controls for the building's fire protection systems located together in one place designated as the fire control centre.

Smoke extraction should be provided from each storey by openable windows or by a mechanical smoke extraction system.

Diesel engine generators or pumps should be enclosed by walls and a roof, of non-combustible construction having a standard of fire resistance of not less than four hours. Any openings in the walls should be protected by a single self-closing 120 minute door.

Oil storage in connection with the foregoing should comply with the fire authority requirements.

Transformer substations and switchrooms containing electrical oil-cooled transformers or oil-filled switchgear with an oil capacity in excess of 250 litres will be required to be housed in a fire-resistant substation area. Cast resin, or dry-type transformers may be a viable option, although initially more expensive.

Conditions will normally be imposed requiring details of the general heating, lighting, electrical and ventilation arrangements to be submitted to Building Control for approval before commencement of the works and such works to be provided and maintained to the satisfaction of Building Contol. Periodic inspections may be made by local Building Control inspectors of the approved heating, lighting, electrical and ventilation installations during installation.

The foregoing Section 20 requirements apply specifically to London but other large cities have their own similar requirements.

5.7 Buried cables

Cables to be installed in ducts or pipes in the ground are not required by BS 7671 to have armouring and/or a metal sheath, if the duct or pipe is sufficiently strong to resist likely mechanical damage. In case of doubt, a cable fulfilling the requirements of Regulation 522.8.10 should be provided. 522.8.10

Protective conductors which are buried directly in the ground must comply with the requirements of Regulation 542.3.1. 542.3.1

All buried cables are required to be marked by cable covers or a suitable marker tape and buried conduits or ducts must also be suitably identified. A warning tape above the duct or conduit run would be suitable. 522.8.10

Cables, conduits and ducts need to be buried deep enough to avoid being damaged by any disturbance of the ground reasonably likely to occur. This could include gardening/horticultural works, excavating for any local buried service pipes, etc. Generally, it is better to locate cables, etc., clear of such possible works.

A depth of burial of less than 0.5 m is usually inadvisable as shallow laid cables may be inadvertently damaged by general gardening, etc. Cables that cannot be buried at a reasonable depth should be specifically protected, e.g. by ducts encased in concrete, or installed along an alternative route.

Before any excavation is undertaken for cable or other works HSE guidance booklet HSG47 *Avoiding danger from underground services* should be studied, as it provides valuable advice on safety aspects.

In areas of agricultural and horticultural premises where vehicles and mobile agricultural machines are operated, cables must be buried in the ground at a depth of at least 0.6 m, with added mechanical protection. Cables in arable or cultivated ground must be buried at a depth of at least 1 m. 705.522

In caravan parks, campgrounds and similar locations, cables must be buried at a depth of at least 0.6 m and, unless having additional mechanical protection, be placed well 708.521.1.1

outside any caravan pitch or away from areas where tent pegs or ground anchors are expected to be hammered into the ground.

514.9 It is important to be able to identify exactly where hidden services are located and accurate records, including drawings, should be made before trenches are backfilled.
522.8.10 Cable routes should be located on drawings by dimensions from some fixed object that is not expected to change for a considerable time (e.g. a building). Road and pathway features may be relatively easily changed. Scale drawings are of little value unless very detailed. Concrete route markers may be installed at changes of direction of cable routes and at regular intervals along routes. It may also be useful to lay cable warning marker tape just below ground level along cable routes so this will be exposed at any future digging well before cables are accessed or disturbed.

The Electricity Safety, Quality and Continuity Regulations 2002 require that all buried cables shall be marked or otherwise protected (Regulation 14(2)).

Many public service undertakings with services buried along the public highway have agreed in the National Joint Utilities Group (NJUG) a specific colour identification system for their service or duct work. Generally, each of the Statutory Undertakings has been allocated a specific colour for their underground services. Details are given in Table 5.2.

In addition, Local Authorities and, in particular, Highway Authorities have allocated themselves colours appropriate for their own installations. These vary from area to area and on occasion may clash with the NJUG colour code. In the interests of safety, Local Authorities should always be asked to supply details before excavation is commenced in their area.

The colour code is not universally applied, as many of the older services were installed before this allocation was developed. Other means of identifying cables include clay or concrete tiles laid above electricity cables, usually with the word 'electricity' embossed into the top surface. The use of coloured plastic tape or board which would have the details of the service printed on its surface is a more recent method.

Further guidance may be obtained from the National Joint Utilities Group or other specifiers (see Appendix L), or any public services supplier.

522.8 Where cables pass through holes in metalwork, brickwork, timber, etc., precautions must be taken to prevent damage to the cables from any sharp edges, or alternatively mechanical protection should be provided, sufficient to prevent abrasion or penetration of the cable.

Notes to Table 5.2:

1. The NJUG code is not retrospective and older installations installed before the code was adopted may not conform to the above colour scheme.
2. Cables installed on private property may not follow the guidelines in this document. The owner concerned should be contacted for information on the colour code applying in these cases.
3. Abandoned ducts are sometimes used for other purposes. Ensure systems are fully identified and traced out before any work is carried out.

▼ **Table 5.2** Recommended colour coding of underground ducts, pipes, cables and marker/warning tapes for public service undertaking (taken from NJUG guidelines on the positioning of underground utilities apparatus for new development sites, Volume 2, December 2007)

Utility	Duct	Pipe	Cable	Marker system	Recommended minimum depths (mm)* Footway/verge	Carriageway
Electricity EHV (Extra High Voltage)	Black	N/A	Black	Yellow with black and red legend or concrete tiles	750–1200	750–1200
Electricity HV (High Voltage)	Black or red tile	N/A	Red	Yellow with black legend	450–600	750
Electricity LV (Low Voltage)	Black or red	N/A	Black or red	Yellow with black legend	450	600
Gas	Yellow	Yellow or yellow with coloured stripes that denote peelable skin pipe of various wall thicknesses	N/A	Yellow with black legend	600 (footway) 750 (verge)	750
Water – non-potable and grey	N/A	Black with green stripes	N/A	N/A	600–900	600–900
Water – firefighting	N/A	Black with red stripes or bands	N/A	N/A	600–900	600–900
Oil/fuel pipelines	N/A	Black	N/A	Various surface markers Marker tape or tiles above red concrete	900 *All work within 3 metres of oil fuel pipelines must receive prior approval*	900 *All work within 3 metres of oil fuel pipelines must receive prior approval*
Sewerage	Black	No distinguishing colour/ material (e.g. ductile iron may be red; PVC may be brown)	N/A	N/A	Variable	Variable
Telecoms	Grey, white, green, black, purple	N/A	Black or light grey	Various	250–350	450–600
Water	Blue or grey	Blue polymer or blue uncoated iron/GRP blue polymer with brown stripe (removable skin revealing white or black pipe)	N/A	Blue or blue/black	750–900 *Depth may vary in accordance with individual company standards*	750–900 *Depth may vary in accordance with individual company standards*
Water pipes for special purposes (e.g. contaminated ground)	N/A	Blue polymer with brown stripes (non-removable skin)	N/A	Blue or blue/black	750–900 *Depth may vary in accordance with individual company standards*	750–900 *Depth may vary in accordance with individual company standards*

* All depths are from the surface level to the crown of the apparatus.

5.8 Sealing and fire stopping

527.2.1 Where a cable, conduit, trunking or busbar trunking system passes through a building wall or floor slab it is necessary to seal the hole around the cable, etc., to the standard of the original wall or slab to prevent the passage of moisture, condensation, vermin or insects, etc. The sealant must be suitable for the cable, etc. and the environment
527.2.7 and should be flexible to allow some small movement of the cable, etc., relative to the
527.2.2 building. Seals should be installed when necessary to maintain the building integrity during construction.

527.2.4 Where the penetration is through any part of the construction identified as having a specified fire performance or fire resistance, the seal must maintain the required fire integrity. The sealing method must be flexible and should be specified by the installation designer in liaison with other design team members such as the architect. The seal must be properly installed through the thickness of the wall or slab and
611.3 inspection of the installation of the seal is required by Regulation 611.3 item (vii). It is unlikely, however, that the electrical installer will be able to determine the suitability of a particular sealing method and specialist advice should be sought.

Conduits, trunking, ducts or busbar trunking systems passing through the building elements have space inside them that can conduct combustion products or other harmful materials (e.g. explosive gases and vapours) and may need to be sealed internally also. Busbar trunking systems are required by BS EN 60439-2 to provide a fire barrier unit for use when the busbar trunking system passes through horizontal or vertical building divisions. This requirement can be met by the method of construction or by internal fire barrier(s) positioned by the manufacturer to correspond to the building arrangement and certified, for a specific time under fire conditions, to ISO 834.
527.2.6 Regulation 527.2.6 states that a non-flame propagating wiring system (see section 5.1) with a maximum internal cross-sectional area of 710 mm^2 (i.e. 32 mm diameter conduit or smaller or 25 x 25 trunking) does not require to be internally sealed for fire protection. It may, however, still require to be sealed for other reasons, such as to prevent condensation.

Sizing of cables 6

6.1 Current-carrying capacity and voltage drop

The Preface to the current rating and voltage drop tables of Appendix 4 of BS 7671 advises how they should be used to calculate the cross-sectional area of a conductor. This will depend, among other factors, on the type of overcurrent protection provided. Guidance Note 6 gives further information on cable sizing. Appx 4

The design current of a circuit should be that drawn by the equipment at the nominal voltage of the supply.

Table 52.3 of BS 7671 considers only the minimum cross-sectional area of conductor that should be used. The actual cross-sectional area chosen depends on a number of factors, which are explained in the Preface to the tables of Appendix 4 of BS 7671. 524.1 Appx 4

Some loads such as motors, transformers and some electronic circuits take large surge (inrush) currents when they are switched on. Due account must be taken of any extra cumulative heating effect due to motor inrush currents or electric braking current where the motors are for intermittent duty with frequent stopping and starting. 552.1.1

6.2 Diversity

Diversity at any point in time is the actual load (or estimated maximum demand) divided by the potential load (or connected load), usually expressed as a percentage. The load that may be taken for the basis of selection of cable and switchgear capacities, allowing the design diversity, must be the maximum foreseeable load figure that will occur during the life of the installation. 311.1

The maximum connected load is usually the simple sum of all the electrical loads that are, or may be, connected to the installation. The electrical load of each item will be the maximum electrical load that item may require, such as the load of an electric motor, unless specific measures have been taken to ensure that, in a group of items with high inertia starting loads, or inrush currents, such loads cannot occur simultaneously. Frequent motor starting will require special consideration.

A safe design would be 100 per cent of the potential (or connected) load but this clearly is not reasonable. This is recognised, for example, in Appendix C of this Guidance Note, where an unlimited number of 13 A socket-outlets may be connected to a single 30 A or 32 A ring final circuit. If there are twenty 13 A socket-outlets (each capable of supplying approximately 3 kW at 230 V) the potential load of the circuit is 60 kW but the maximum permissible for a 30 A circuit is 6.9 kW, a diversity of 11.5 per cent. In practice, the actual load on a dwelling ring final circuit is usually well below 6.9 kW, except perhaps in a kitchen or utility room.

The economic design of an electrical installation will almost always mean that diversity has to be allowed. However, where there is doubt as to the factors to be used, an adequate margin for safety should be allowed and consideration should also be given to the possible future growth of the maximum demand of the installation.

433.1 Section 433 requires circuits to be so designed that small overloads of long duration are unlikely to occur.

Further information is contained in Appendix C.2 'Final circuits using socket-outlets and fused connection units complying with BS 1363'.

For a guide towards the estimation of diversity see Appendix H.

The loads of ring final circuits should be shared equally around the ring. A cluster of current-using equipment near to the distribution board can overload one leg of the ring.

6.3 Neutral conductors

Consideration must be given to the cross-sectional areas of neutral conductors. In a
524.2 single-phase circuit the neutral conductor must have at least the same cross-sectional
524.3 area as the line conductor. In circuits for discharge lighting, a reduced neutral conductor is not permitted, where the harmonic content of the line currents is greater than 10 per cent of the fundamental current. However, Regulation 524.3 allows the possibility of a reduced size neutral in polyphase circuits where the neutral conductor will not
523.1 be overloaded (523.1). Even with balanced three-phase conditions, there can be high percentages of third harmonic currents which add arithmetically in the neutral conductor and when assessing the maximum likely neutral current, allowance needs
431.2.3 to be made for any unbalanced loading or harmonics. Overcurrent protection shall be provided for the neutral conductor in a polyphase circuit where the harmonic content of the line currents is such that the current in the neutral conductor is reasonably expected to exceed that in the line conductors.

Where the circuit is for discharge lighting, the design current, in the absence of more exact information, can be taken as:

$$I_b = \frac{1.8 \text{ x lamp rated wattage}}{\text{nominal voltage of circuit}} \text{ amperes}$$

The multiplier (1.8) is based on the assumption that the circuit is corrected to a power factor of not less than 0.85 lagging and it takes into account controlgear losses and harmonic currents. If corrected discharge lighting is used, the power factor will be typically 0.85 lag but uncorrected discharge lamps may vary between 0.5 lead and 0.3 lag.

For high-frequency circuits, Lighting Industry Federation (LIF) Technical Statement 21 should be consulted.

High neutral currents may be encountered with information technology equipment using switch-mode power supplies. In such equipment the mains input is rectified, the d.c. output being fed to a capacitor which in turn supplies the switch-mode regulator.

It has been suggested that the rms current in the neutral conductor of a three-phase mains circuit can be 1.73 times the rms current in the line conductors.

Thus, when designing three-phase circuits for equipment incorporating switch-mode power supplies the designer may need to determine from the equipment manufacturer the rms current taken by the equipment and the expected neutral current and its harmonic content. It may sometimes be necessary to install a neutral conductor of a larger size than the associated line conductors; however, it is preferable to attenuate the harmonic distortion by the use of active filters or similar devices.

6.4 Voltage drop in consumers' installations

The prime requirement of this section is that the voltage at the terminals of any fixed current-using equipment should be such that it is greater than the lower limit allowed by the product standard for that equipment. In practical terms it means that the voltage drop should not be such as to impair the safe and effective working of the equipment. 525.1 525.2

In the absence of a product standard, the manufacturer's instructions should be followed. 525.4

The voltage drop between the origin of an installation and any load point should not be greater than the values in Table 6.1, i.e. Table 12A in Appendix 12 of BS 7671, which shows values in percentage terms of permitted voltage drop expressed with respect to the value of the nominal voltage of the installation. 525.3 Appx 12

The calculated voltage drop should include any effects due to harmonic currents.

	Lighting	Other uses
1 Low voltage installations supplied directly from public low voltage distribution system	3%	5%
2 Low voltage installation supplied from private LV supply*	6%	8%

▼ **Table 6.1** Maximum values of voltage drop (Table 12A of BS 7671)

* The voltage drop within each final circuit should not exceed the values given in **1**.

▶ Where the wiring systems of the installation are longer than 100 m, the voltage drops indicated above may be increased by 0.005% per metre of the wiring system beyond 100 m, without this increase being greater than 0.5%.

▶ The voltage drop is determined from the demand by the current-using equipment, applying diversity factors where applicable, or from the value of the design current of the circuit.

Notes:

1 A greater voltage drop may be acceptable for a motor circuit during starting and for other equipment with a high inrush current provided that in both cases it is ensured that the voltage variations remain within the limits specified in the relevant equipment standard.

2 The following temporary conditions are excluded:

- voltage transients
- voltage variation due to abnormal operation

Voltage drops may be determined from Appendix 4.

The Electricity Safety, Quality and Continuity Regulations 2002 allow a supply variation of +10% to -6%. When designing large installations with a high voltage supply, the designer can keep the voltage drop on the main distribution circuit to within, say, 10% and allow, say, 4% on final circuits to achieve a practical design.

For intermediate installations it may be necessary to check the actual voltage range at the supply and by calculation to determine the actual voltage drop that will be acceptable to keep the voltage at the final sub board within limits.

When deciding on the voltage drop limits to be applied, the designer must bear in mind that a 5% reduction in voltage may result in a 10% reduction in power output and an even larger percentage drop in the light output of incandescent lamps. Data on the performance and life expectancy of incandescent lamps can be obtained from manufacturers.

Other influences 7

7.1 Electrical connections

Connections between cables and other equipment need to be selected with care. There is a risk of corrosion between dissimilar metals, depending upon the environment. The manufacturer's recommendations must be observed and the Standard followed. The power rating and the physical capacity of terminals must be adequate, as must their physical strength, when terminating larger conductors. 522.5 522.8.5

Thermosetting insulated cables to BS 5467 and thermoplastic cables to BS 6004 can operate at higher conductor temperatures than other thermoplastic insulated cables. The higher conductor temperature results in an increased current-carrying capacity (10–20%) and consequently higher voltage drop values.

The use of high temperature cable requires careful selection of termination systems and accessories and switchgear which are appropriate to the operating temperature of the cable. Alternatively, where high temperature cable is to be used, it should be derated (operating current I_z reduced) so as not to allow the conductor temperature to exceed the maximum operating temperature of the equipment to which it is to be connected. Conversely, high temperature cables may be a necessity where the heat of the device may be excessive. Thermoplastic or thermosetting conductors used in this way may need their insulation locally replaced or supplemented by an insulation with a more suitable temperature rating (e.g. at an enclosed luminaire). Where it is necessary to adjust the cross-sectional area of a cable at the point of termination this should be done by the use of a recognised technique. 512.1.2

Any enclosure must have suitable mechanical and fire resistance properties. Unless exempt, joints must remain accessible for inspection and maintenance. 421.6 526.3, 526.4 543.3.3

Account must be taken of any mechanical damage and vibration likely to occur. The method of connection adopted should not impose any significant mechanical strain on the connections, nor should there be any mechanical damage to the cable conductors. 526.2 526.6

Cable glands, clamps and compression-type joints should retain securely all the strands of the conductor. Every compression joint should be of a type which has been the subject of a test certificate as described in BS EN 61238-1:2003, and the appropriate tools and methods specified by the manufacturer of the joint connectors should be used. A termination or joint in an insulated conductor, other than a 4 mm^2 or greater protective conductor, should be made in an accessory or luminaire complying with the appropriate British Standard. Where this is not practicable, it should be enclosed in material designated non-combustible when tested to BS 476-4 or complying with 'P' to BS 476-12 or the relevant requirements of BS 6458, Section 2.1. Such an enclosure may be formed by part of an accessory and/or luminaire and a part of the building structure. 526.1

421.7 All connections, including those of ELV 12 V luminaires, must be enclosed in accordance
526.5 with the requirements of Regulations 421.7 and 526.5. The use of unenclosed or taped-up terminal blocks or strips, which themselves may not be tested to any recognised standard, does not comply with the requirements of the Regulations. Connections must be enclosed within a properly constructed junction box or enclosure, either to the relevant British Standard (e.g. BS 4662 or BS EN 60670) or made to an equivalent flame propagation which passes the glow-wire test of BS EN 60695-2-1/0, 1/3, or be in an enclosure formed from a suitable building material or formed within the building structure. Such building material must be non-combustible, or have suitable ignitability
526.9 characteristics when tested as specified in BS 476. All insulated, unsheathed cable, such as cable ends stripped back, must be contained within the enclosure.

Terminations of mineral insulated cables should be provided with sleeves having a temperature rating similar to that of the seals. Cable glands should securely retain without damage the outer sheath or armour of the cables. Mechanical cable glands for thermosetting and plastics insulated cables should comply with BS 6121 where appropriate.

553.2 Appropriate cable couplers should be used for connecting together lengths of flexible cable or flexible cord. Generally cable couplers shall be non-reversible and to an
704.511.1 appropriate Standard. On construction sites, where the requirements apply, every plug and socket-outlet of rating 16 A or greater must comply with BS EN 60309-2 (BS 4343).

Conduit should be free from burrs and swarf and all ends should be reamed to obviate damage to cables. Substantial boxes of ample capacity should be provided at every junction involving a cable connection in a conduit system. The designer/installer should be aware that large cables need a greater factor of space for installation and removal.

537.2.1.7 Regulation 537.2.1.7 requires a provision for disconnecting the neutral conductor for isolation or testing purposes. A suitable joint will be adequate for smaller cables but a
537.2.2.4 bolted link could be used for larger conductors due to the physical effort necessary to move the cables for disconnection/reconnection.

7.2 Cables in contact with thermal insulation

523.7 Where a cable is to be run in a space in which thermal insulation is likely to be placed, the cable should, wherever practicable, be fixed in a position such that it will not be in contact with or covered by the thermal insulation. Where fixing in such a position is impracticable (see Figure 7.1) the cross-sectional area of the cable must be appropriately increased.

▼ **Figure 7.1**
Cables in thermal insulation

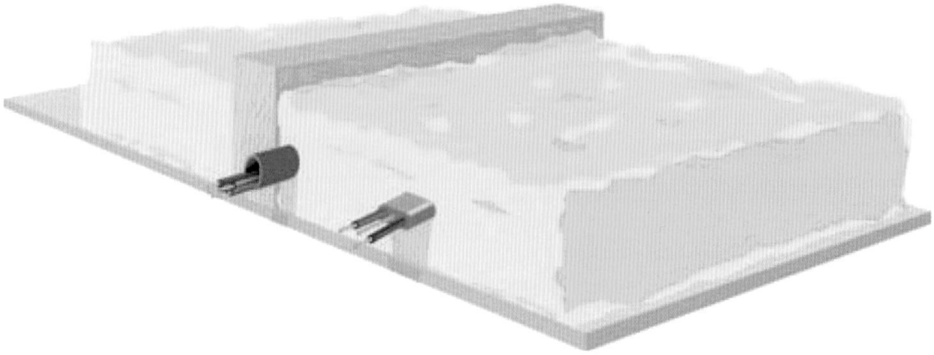

For a cable installed in a thermally insulated wall, the cable being in contact with a thermally conductive surface on one side, current-carrying capacities are tabulated in Appendix 4 of BS 7671, Method A being the appropriate reference method. For flat twin and earth cables run in stud walls or above plasterboard ceilings, Methods 100–103 are applicable. Appx 4

For a single cable likely to be totally surrounded by thermally insulating material over a length of more than 0.5 m, the current-carrying capacity must be taken, in the absence of more precise information, as 0.5 times the current-carrying capacity for that cable clipped direct to a surface and open (Reference Method C).

Where a cable is totally surrounded by thermal insulation for less than 0.5 m the current-carrying capacity of the cable must be reduced appropriately depending on the size of cable, length in insulation and thermal properties of the insulation. The derating factors in Table 7.1 are appropriate to conductor sizes up to 10 mm^2 in thermal insulation having a thermal conductivity greater than 0.04 $Wm^{-1}K^{-1}$. Table 52.2

Length in insulation (mm)	Derating factor
50	0.88
100	0.78
200	0.63
400	0.51

▼ **Table 7.1**
Cables surrounded by thermal insulation (Table 52.2 of BS 7671)

The above table is based upon normal cavity insulation such as vermiculite granules. More precise data can be obtained from ERA Report 85-0111 'The temperature rise of cables passing through short lengths of thermal insulation'.

The current-carrying capacities are the same for cables installed in the following manner (Reference Method C): Table 4A2

1 Sheathed cables clipped direct (Installation Method 20)
2 Sheathed cables embedded directly in masonry, plaster and the like (Installation Methods 57 and 58).

For cables in conduit in a thermally insulating wall or above a thermally insulating ceiling and where the conduit is in contact with a thermally conductive surface on one side (Installation Method 2), the current-carrying capacities are lower than for Reference Method C, i.e. Method A is applicable.

The following notes consider the effect of such derating on various typical circuits.

For domestic lighting circuits, the load does not normally exceed 6 A. The smallest cable used will be 1 mm^2 with a current-carrying capacity for Method C of 16 A (Table 4D5, column 6). If totally surrounded by thermal insulation for 0.5 m or a greater length the current-carrying capacity can therefore be taken as 8 A, which would be acceptable with a 6 A overcurrent protective device. Appx 4 433.1.1(ii)

A cable sometimes used for cooker circuits is 6 mm^2, which has a current-carrying capacity of 47 A when clipped direct, reducing to 23.5 A if it is installed in thermal insulation. A 30 A or 32 A overcurrent device is usually employed. The current taken by a cooker varies over a wide range and theoretical loads in excess of 23.5 A can be expected. This would lead to overheating of the cable if continuously carrying such
433.1.1 (ii) current. Hence, a larger cable may be necessary. Also, the nominal current (I_n) of a protective device must not exceed the lowest of the current-carrying capacities of any of the circuit conductors (I_z).

The loading of cables feeding socket-outlets is even more varied and depends on the type of circuit used, i.e. ring or radial, the number of socket-outlets fed from the circuit and the type of load. For example, 2.5 mm^2 thermoplastic insulated and sheathed cables used in a multi-socket-outlet radial circuit protected by a 20 A overcurrent protective device have a 'clipped direct' current-carrying capacity of 27 A (Table 4D2A, Column 6), reducing to 13.5 A if the cable is embedded in thermal insulation.

When clipped direct the cable could supply up to 6.2 kW without overheating, though because, in this example, the circuit is protected by a 20 A device it should not be loaded to more than 4.6 kW in any event.

Loading as great as this in domestic installations is infrequently encountered and much more common are small loads such as standard lamps and television sets, up to about 1 kW, i.e. 4 A in total.

However, if the cable is later embedded in thermal insulation, the reduced current-carrying capacity of 13.5 A would be exceeded if a load of 3.2 kW or greater were connected and overheating would be a possibility.

Thus, the introduction of thermal insulation into cavities where cables are already installed will create the possibility of overheating the cable. The extent of the risk will depend on the type of thermal insulation used and the total loading of the cables.

7.3 Mutual or individual deterioration

522.5.3 BS 7671 requires that materials liable to cause mutual or individual deterioration or hazardous degradation shall not be placed in contact with each other. One popular type of thermal insulation for cavity walls is urea formaldehyde foam and there should be no adverse action between this material and thermoplastic.

Other materials may be used for cavity insulation but the supplier does not always disclose their chemical composition. Before allowing any thermal insulation materials to come into contact with cable insulation or sheath materials the thermal insulation supplier should be asked to confirm in writing that there will be no adverse effect on the cable insulation or sheath. Where there is doubt, an inert barrier should be inserted between the cable and the thermal insulation.

Expanded polystyrene sheets or granules are used in construction (e.g. insulating lofts and supporting floors) and may be used in cavity walls. If this material comes into contact with thermoplastic cable sheathing, some plasticiser will migrate from the PVC to the polystyrene. The thermoplastic will become less flexible and sticky on the surface and the polystyrene will become soft and shrink away from the thermoplastic sheathed cable if possible. Such contact between thermoplastic and polystyrene should be avoided. These comments do not apply to unplasticised PVC conduit and trunking systems.

As only the plasticiser is removed from the sheath, it is not expected that the insulation resistance of the cable will be affected if the cable is not disturbed but this cannot be guaranteed. The cable should be supported away from the polystyrene and a test should be carried out. The cable condition should also be monitored over time in regular periodic inspections. There is no cure for this migration other than replacement of the cable.

Similar remarks apply to some fittings which are in contact with cables. In addition to polystyrene and expanded polystyrene, acrylonitrile-butadiene-styrene (ABS) and polycarbonate are also affected. Nylon, polyester, polyethylene, polypropylene, rigid PVC and most thermosetting plastics are little affected. Natural rubber grommets can become softened but synthetic rubber and PVC grommets are satisfactory.

Contrary to popular belief, plastic insulated and sheathed cables are not suitable for continuous immersion in water, such as in pools and fountains, as they will absorb an amount of water over time and insulation failure will occur. Manufacturers' advice should be taken for applications involving continued immersion of cables.

7.4 Proximity of wiring systems to other services

Electrical and all other services shall be protected from any harmful mutual effects foreseen as likely under conditions of normal service. 515.1 Section 528

Section 528 contains a reference to two voltage bands, equating approximately to ELV and LV. Care must be taken to make sure that only compatible circuits are enclosed in the same conduit or trunking, unless the circuit conductors are insulated for the highest voltage present, or other precautions are taken, such as segregation by a separate compartment in an insulated cable system or by an earthed metallic screen.

For safety circuits such as emergency lighting and fire alarms, covered by BS 5266 and BS 5839 respectively, the requirements of the British Standards must be met with respect to segregation. For telecommunications see BS 6701 and Appendix I. 560.7.7

Where malfunction may occur (e.g. crosstalk with a communication system) all conductors, including the protective conductor, must be correctly routed and adequate separation provided from other cables (see Appendix I also). A particular form of harmful effect may occur when an electrical installation shares the space occupied by an audio frequency induction loop system – this loop system enables a hearing aid with a telecoil to pick up audio signals from the loop. Under these circumstances, if line and neutral conductors or switch feeds and switch wires are not run close together, 50 Hz interference (and its harmonics) may be picked up by the telecoil of the hearing aid.

Conventional two-way switching systems can often produce inadvertent 50 Hz radiating loops but this can be reduced by arranging the circuits as shown in Figure 7.2.

All the conductors of a circuit should generally follow the same route. Live cables of the same circuit may cause overheating if they enter a ferromagnetic enclosure through different openings. 521.5.2

The EMC Regulations require that attention is paid to electrical equipment and installations to limit their electrical interference from other systems (immunity) and their electrical interference with other systems (emission). 515.3

▼ **Figure 7.2**
Circuit for reducing interference with induction loop

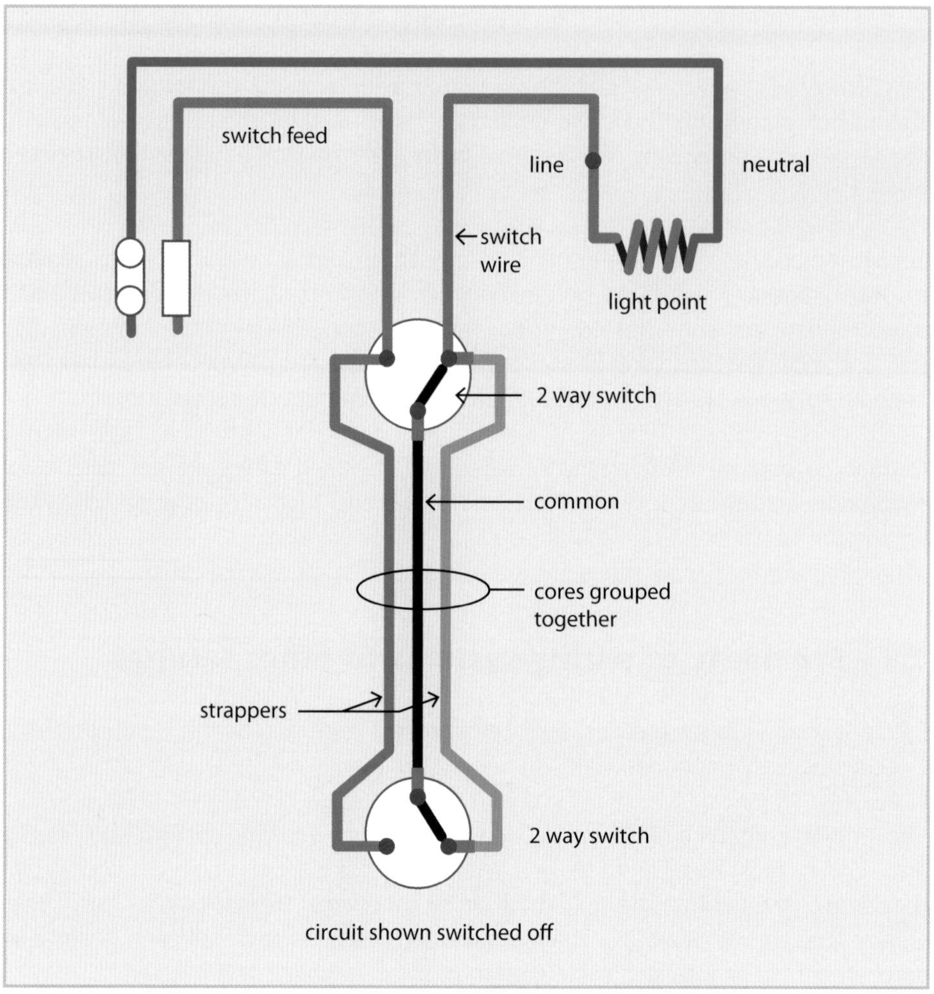

Electrical power systems will usually have immunity from interference from other systems but may cause emission interference with sensitive electronic systems. In addition, harmonic generation and interference will need to be considered. Electronic systems such as data systems, fire alarms and public address will give the most concern but detailed information and advice on emission and immunity is not generally available for systems and each installation must be separately considered by experts and any manufacturer's advice acted upon.

Controlgear and factory-made equipment which is certified or declared to comply with relevant emission and immunity standards can be expected to do so but on site a cable installation may pick up or induce interference unless it is adequately separated or protected. The data in Appendix I can be utilized in the absence of other guidance but no firm data is available to guarantee these figures. The Electrical Contractors' Association (ECA) publication *Recommended cable separations to achieve electromagnetic compatibility (EMC) in buildings* (reproduced in part in Appendix I) gives some advice which may be useful.

7.5 Plasticiser migration from PVC insulation

Installers may occasionally encounter a sticky blue/green deposit in equipment or switch and socket-outlet boxes in older installations wired in thermoplastic insulated cables.

Early manufacture of PVC insulation materials utilized a different formulation from that which is now employed. Thermal cycling of the conductor due to changes in load and conductor temperature drew the thermodynamically unstable insulation plasticiser to the conductor, where, on vertical runs of cable over time, the plasticiser liquid worked down the conductor surface to become a sticky blue/green liquid in the accessory box. The material is not corrosive and can be cleaned away with methylated spirit. The blue/green colouring comes from traces of copper absorbed by the liquid. There is no cure for this plasticiser migration, other than replacement of the cable.

As only the plasticiser is removed from the insulation, it is not expected that the insulation resistance of the cable will be affected if the cable is not disturbed but this cannot be guaranteed and a test should be carried out. The cable condition should also be monitored over time in regular periodic inspections.

Reformulation of PVC compounds in the mid to late 1970s has removed this problem from modern PVC insulated cables.

Migration also happens when bitumen is in contact with PVC. The bitumen absorbs some plasticiser from the PVC. While the amount of loss is insufficient to have much effect on the properties of the PVC, it is enough to cause considerable reduction in the viscosity of the bitumen, which may become so fluid that it can run.

Installation of equipment

8

8.1 Earthing requirements for the installation of equipment having high protective conductor currents

8.1.1 Scope

The requirements of this regulation group apply to: 543.7

1. equipment between the final circuit wiring and current-using equipment where the protective conductor current exceeds 3.5 mA
2. final circuits and distribution circuits where the accumulated protective conductor current is expected to exceed 10 mA.

The use of information technology (IT) equipment, which has low electrical loading but produces relatively high protective conductor currents, often requires special precautions to be taken. A final circuit can accept many items of IT equipment and remain within the rating of the circuit but the protective conductor currents resulting may be relatively very high.

Modern equipment, such as electronic ballasts in high-frequency fluorescent luminaires or variable-speed drives, is known to create high currents in the protective conductor, hence the precautions of Regulation 543.7 of BS 7671 must be followed.

Levels of permitted protective conductor currents for certain types of equipment are given in Appendix J. The information is taken from relevant British Standards. Where precise information is required the relevant standards should be consulted.

The levels of expected protective conductor current will vary depending on the type of electrical equipment. *BS EN 60335-1 Household and similar electrical appliances – Safety – Part 1: General requirements* states the fundamental conditions to which items of electrical equipment are manufactured and, hence, should perform when used under normal operating conditions.

Clause 13.2 of BS EN 60335-1 states that after the appliance has been operated for a duration corresponding to the most unfavourable conditions of normal use, the leakage current (protective conductor current) should not exceed the values shown in Table J.1.

Regulation 543.7 contains additional earthing requirements for the installation of equipment having high protective conductor currents. These particular requirements complement the protection described in Section 531 and in Chapters 41 and 54 of BS 7671. Information technology and other electronic equipment usually incorporates filters on the mains conductors. In Class I equipment these filters include capacitors connected between the supply conductors and frame and this results in a standing protective conductor current.

For information technology equipment intended for connection to the supply via a 13 A BS 1363 plug, the maximum allowable standing protective conductor current is 3.5 mA as specified in BS EN 60950. However, modern equipment usually has leakages significantly less than this, perhaps 1 mA to 1.5 mA per item.

With desktop equipment, the use of a number of items of portable equipment connected to a single socket-outlet or a ring or radial final circuit can result in a cumulative protective conductor current of several milliamperes. This current may be high enough to present danger if the protective conductors become open circuit and, also, may operate an RCD. It should be remembered that an RCD to BS 4293, BS EN 61008 or BS EN 61009 is allowed to operate at any value of residual current between 50 and 100 per cent of its rated residual operating current. Therefore, unwanted tripping could occur under normal running conditions. This is even more likely if there is a surge of protective conductor current when switch-on occurs near the peak value of the supply voltage.

Regulation 543.7 therefore requires that the circuit arrangement must be such that the residual current which may be expected to occur, including switch-on surges, will not trip the device.

Information technology equipment connected via BS EN 60309-2 (BS 4343) industrial plugs and socket-outlets is permitted to have higher protective conductor currents. However, there is a limit of 10 mA above which the special installation requirements detailed in Regulation 543.7 of BS 7671 apply.

These special requirements are aimed at providing a high-integrity protective conductor. Should the protective conductor of the circuit feeding the equipment fail, the protective conductor current could then pass to earth through a person touching the equipment. As with all installations, the high-integrity protective conductor installation must be periodically inspected and tested to ensure its integrity is maintained (see Guidance Note 3 for further details).

8.1.2 The risks

The risk associated with final circuits with high protective conductor currents is that resulting from discontinuity of the protective conductor, and serious shocks can be received from accessible conductive parts which are not connected to the main earthing terminal of the installation.

This shock risk is extended to all the items of equipment on that particular circuit, whether they individually have high protective conductor currents or not. The more equipment that is connected to a circuit, the wider the risk is spread and the greater the hazard.

IEC Publication 479-1 *The effects of current passing through the human body* (BS PD 6519-1) advises that, for body currents less than 10 mA, there are usually no harmful physiological effects.

From Table 4 of IEC 479-1, protective conductor currents exceeding 10 mA can have harmful effects and precautions need to be taken.

543.7.1.2 The requirements of Regulation 543.7 are intended to increase the reliability of the
543.7.1.3 connection of protective conductors to equipment and to earth, when the circuit
543.7.1.4 protective conductor current exceeds 10 mA. This normally requires duplication of the protective conductor or an increase in its size in the final circuit. The increased size is not to allow for thermal effects of the protective conductor currents, which are

insignificant, but to provide for greater mechanical strength and, it is intended, a more reliable connection with the means of earthing. Duplication of a protective conductor, each terminated independently, is likely to be more effective than an increase in size.

8.1.3 Equipment

Note that BS EN 60950, the standard for the safety of information technology equipment, including electrical business equipment, requires equipment with a protective conductor current exceeding 3.5 mA to have internal protective conductor cross-sectional areas not less than 1.0 mm² and also requires a label bearing the following or similar wording fixed adjacent to the equipment primary power connection:

HIGH LEAKAGE CURRENT **Earth connection essential** **before connecting the supply**

or

WARNING HIGH TOUCH CURRENTS **Earth connection essential** **before connecting the supply**

Regulation 543.7.1.1 requires a single item of equipment, if the protective conductor current exceeds 3.5 mA but does not exceed 10 mA, to be permanently connected or connected by a plug and socket complying with BS EN 60309-2. If the protective conductor current of a single item of equipment exceeds 10 mA the requirements of Regulations 543.7.1.2 and 543.7.1.3 for a high-integrity protective earth connection must be met. 543.7.1.1 543.7.1.2 543.7.1.3

8.1.4 Labelling

Distribution boards supplying circuits with high protective conductor currents must be labelled accordingly so that persons working on the boards can ensure they maintain the protective precautions taken (see example of labelling in Figure 8.4). 543.7.1.5

8.1.5 Ring circuits

Ring circuits, by their design, provide duplication of all conductors, including the protective conductor unless this is formed by metallic conduit and/or trunking. Regulation 543.7.1.4 requires that each protective conductor is terminated separately at each connection point; the regulation also requires that each wiring accessory has two separate earth terminals. At socket-outlets, this is a very easy requirement to meet – simply select accessories with dual earth terminals (see Figures 8.1 and 8.2).

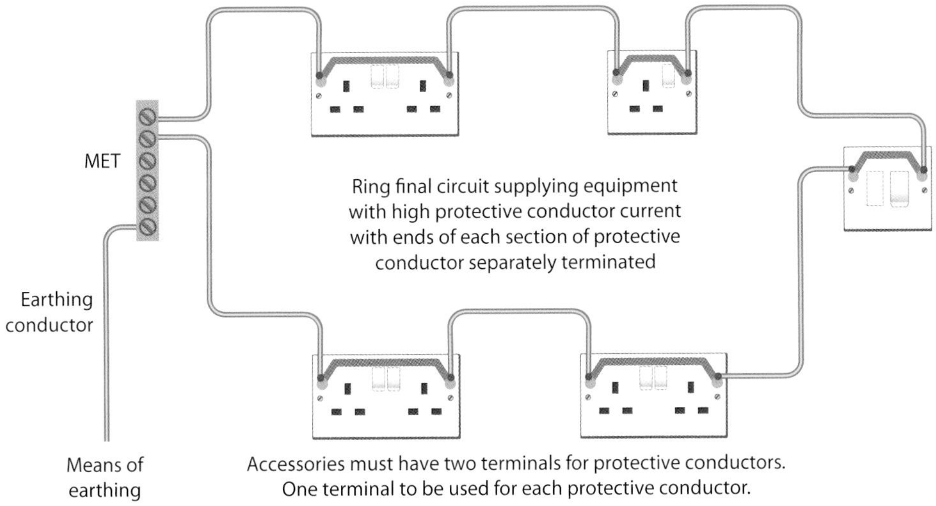

▼ **Figure 8.1** Ring final circuit supplying socket-outlets where the protective conductor current is likely to exceed 10 mA

▼ **Figure 8.2**
Accessory with two terminals for protective conductors

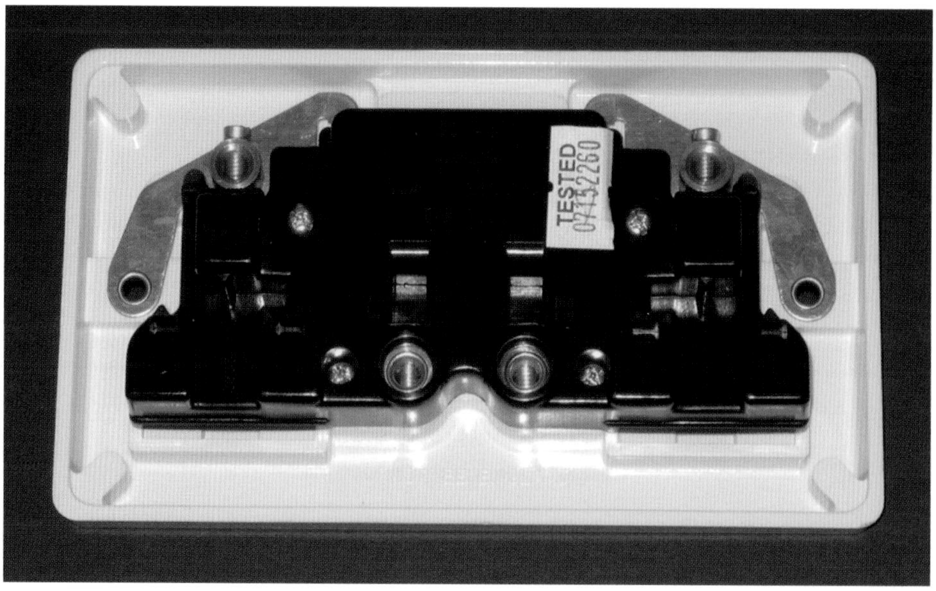

The requirement to terminate duplicate protective conductors separately can pose a problem at the earth bar of the distribution board, i.e. there may not be sufficient terminals to do so. Some large distribution boards may be provided with many spare terminals on the earth bar, while some manufacturers may supply an accessory kit to add extra ways. Where this is not an option, the following example is one method of meeting the requirements.

An electrical installation has a six-way distribution board/consumer unit and consists of six circuits:

DB 1	
Circuit	**Description**
1	Ring final circuit with high protective conductor current
2	Radial circuit with high protective conductor current
3	Ring final circuit
4	Radial circuit
5	Radial circuit
6	Radial circuit

543.7.1.4 Regulation 543.7.1.4 requires that where two protective conductors are used on circuits with high protective conductor currents, they are to be terminated into separate ways on the earth bar.

By installing circuits 1 and 2 across two ways of the earth bar (see Figure 8.3), in addition to marking each conductor with the correct circuit identification ferrule, there is no ambiguity as to the function of each conductor.

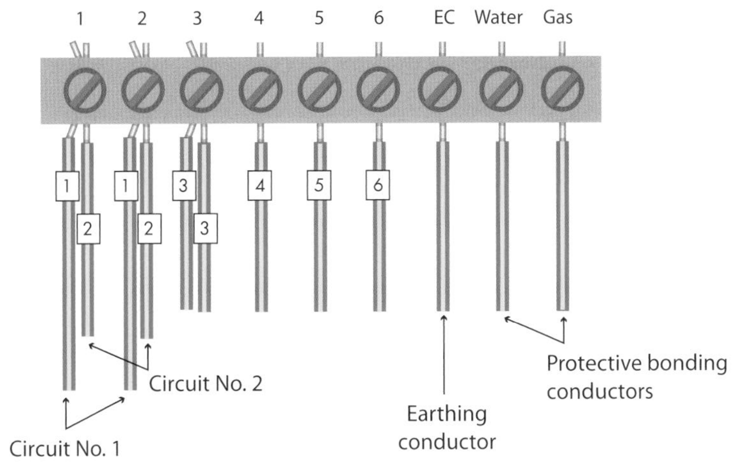

Figure 8.3 Termination of circuits with high protective conductor currents at the earth bar

Regulation 543.7.1.5 requires that information is provided at the distribution board indicating those circuits having a high protective conductor current. This information could be displayed using the format shown in Figure 8.4. 543.7.1.5

DB1
The following have high protective conductor current:
1 Ring final circuit (sales office)
2 Radial final circuit (packaging/dispatch)

Figure 8.4 Information in accordance with Regulation 543.7.1.5 indicating those circuits with high protective conductor current

8.1.6 Radial circuits

Radial circuits supplying sockets where the protective conductor current is expected to exceed 10 mA are required to have a high-integrity protective conductor connection. This can often be most effectively provided by a separate duplicate protective conductor connecting the last socket directly back to the distribution board, as shown in Figure 8.5. This will provide a duplicate connection for each socket on the circuit.

The following requirements still apply:

1. All socket-outlets must have two protective conductor terminals (see Figure 8.2), one for each protective conductor
2. The duplicate protective conductors must be separately connected at the distribution board (see Figure 8.3).

Note: To reduce interference effects, the duplicate protective conductor should be run in close proximity to the other conductors of the circuit.

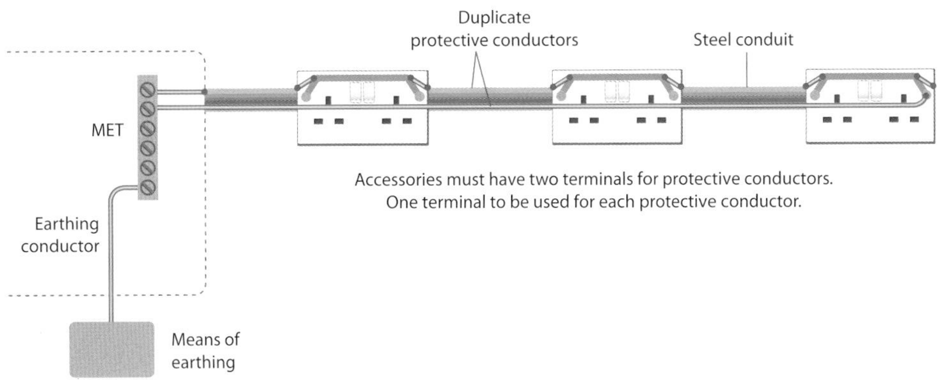

Figure 8.5 Radial circuit supplying socket-outlets (total protective conductor current exceeding 10 mA), with duplicate protective conductor

8.1.7 Busbar systems

Busbar systems are often adopted by designers for supplying IT equipment. These may be radial busbars with tee-offs to individual sockets or floor-boxes. The main PE busbar
543.7.1.3 will need to meet one or more of the requirements of Regulation 543.7.1.3:

- A cross-sectional area not less than 10 mm^2
- A cross-sectional area not less than 4 mm^2 with mechanical protection
- Duplicate protective conductors, each of cross-sectional area sufficient to meet the requirements of Section 543.

Tee-offs from a busbar system

If the protective conductor current is expected to be less than 3.5 mA, there is no need for duplication of the protective conductor from the radial busbar to the socket.

Where the protective conductor current is likely to exceed 10 mA then a single copper protective conductor having a cross-sectional area not less than 4 mm^2 and enclosed to provide additional mechanical protection can be used.

It is necessary for the designer to confirm that the disconnection time of the protective device feeding the busbar will provide adiabatic fault protection to the final conductors.

Overload protection would not be required where the final connection is to a single socket-outlet rated at 13 A, regardless of the size of the protective device for the circuit. However, many floor-box style installations may have up to six socket-outlets per floor-box or branch of the radial circuit and the designer must take this into account when designing such an installation, particularly where the busbar system is protected by a device greater than 32 A.

More information on overload protection can be found in Chapter 3 of this publication and in Guidance Note 6: *Protection Against Overcurrent*.

8.1.8 RCDs and high protective conductor currents

411.3.3 To meet the requirements of additional protection, BS 7671 requires that socket-
415.1.1 outlets not under the supervision of skilled or instructed persons are to be protected
314 by an RCD with a rating not exceeding 30 mA. Where many items of equipment creating high protective conductor currents are protected by the same RCD, unwanted tripping may occur. It is therefore necessary to design circuits with such equipment correctly to ensure that the RCD does not trip unless an unexpected residual current causes it to do so.

In order to test the effectiveness of RCDs after installation, a series of tests may be applied to check that they are within specification and operate satisfactorily. This test sequence will be in addition to proving that the test button is operational. The effectiveness of the test button should be checked after the test sequence.

For each of the tests, readings should be taken on both positive and negative half-cycles and the longer operating time recorded. See Guidance Note 3 and the *On-Site Guide* for more information.

Each type of RCD, e.g. General purpose RCDs to BS 4293, General purpose RCDs to BS EN 61008, RCBOs to BS EN 61009 or RCD-protected socket-outlets to BS 7288, is initially tested by injecting a current which is 50 per cent of the rated tripping current, i.e. 15 mA; under this condition, the RCD must not operate.

The RCD is then tested by injecting a current which is 100 per cent of the rated tripping current, i.e. 30 mA; under this condition, the RCD must operate within a specified time.

As no test is made between 15 mA and 30 mA, it is reasonable to expect that some RCDs will operate when a residual current within those parameters is flowing. Therefore, designers should be aware of all possibilities when designing circuits. It may be prudent to limit the designed protective conductor current to 0.5 $I_{\Delta n}$ (i.e. 15 mA when the circuit is protected by a 30 mA device) and consider surge currents when all items of equipment are energized simultaneously.

8.2 Water heating

It is particularly important that manufacturers' instructions are followed when installing electrode water heaters or boilers, to isolate simultaneously every cable feeding an electrode and to provide overcurrent protection. Similarly, the installation of water heaters having immersed and uninsulated heating elements must follow makers' instructions and be correctly bonded. The metal parts of the water heater (e.g. taps, metal-to-metal joints, covers) must be bonded to the incoming water pipe and the pipe connected to the main earthing terminal with a dedicated protective conductor independent of the circuit protective conductor. The heater must be permanently connected through a double-pole switch and not by means of a plug and socket-outlet (see also Guidance Note 2).

554.1 554.3 554.3.2 554.3.3

8.3 Safety services

Supplies for safety circuits must comply with Chapter 56. The design and construction of any other circuit supplying a safety service shall be such that a fault or change in the other circuit does not affect the correct functioning of the safety service. The routes of cables should not render the safety service liable to abnormal fire risk. Such safety services would include:

560.6 560.7.1 560.7.2 560.1

1 fire alarms
2 emergency lighting
3 designated lifts for use in emergency.

8.4 Other equipment

Chapter 55 covers requirements for accessories and other equipment and most of the requirements are reasonably straightforward. Specific requirements are given for typical equipment and accessories. Attention is drawn to Regulation 559.6.1.1 which allows a choice of accessories to terminate a lighting point.

559.6.1.1

In domestic installations, use of a device for connecting luminaires (DCL) in accordance with BS EN 61995 is recommended, since the interface for this device is harmonized throughout Europe. Additionally, installation of a box at a lighting point, to accommodate the connections, the socket-outlet part of the DCL and the support of a luminaire, has been introduced to overcome the non-regulated but somewhat common and unsafe practice of pushing connections up into a ceiling void. Use of a DCL is intended to permit replacement, cleaning and decoration without the need for access to fixed wiring.

Luminaire supporting couplers (LSCs) or plug and socket-outlet arrangements are alternatives to the harmonized DCL system, each being used to facilitate the removal of luminaires for repair or rearrangement of the lighting layout. Item (v) of Regulation 559.6.1.1 can include any type of socket-outlet for convenience but the provision of a socket-outlet with a higher current rating than the design current of the circuit may require local labelling to stop it being utilized for other purposes.

The installation of DCLs, LSCs or other socket-outlet systems allows an installation to be completed, tested and energized before luminaires are selected and installed. Luminaires can then be fitted (perhaps after decorating) without interference with the installation and consequent retesting.

Many modern lighting installations are controlled by sophisticated management systems with automatic sensors and software control.

There may not yet be a suitable British Standard for the hardware of such systems but the wiring and the selection and erection of the equipment and luminaire connections must still comply with BS 7671.

Accessories must comply with the relevant British Standards and care must be exercised to ensure damage does not occur during installation, operation or maintenance.
512.1 Accessories must be of a type and IP rating appropriate to their location and should not be installed in locations where they may suffer impact, wet or dampness or corrosion
512.2 unless they are of a design specifically for the application. Regulation 512.2 applies. This regulation applies to socket-outlets near sinks which must be of a type resistant to water ingress if they are likely to be splashed (see also Appendix C.6.2).

512.1.2 Care must be taken to ensure that the conductor operating temperature does not allow the permitted temperature rise of the accessory or equipment to be exceeded. Thermosetting insulated cables to BS 5467 or BS 7211 operate at a conductor temperature of 90 °C when carrying their rated current and are not suitable for connection to accessories or devices with a maximum operating temperature of 70 °C unless the cables are derated.

559.6.1.7 B15 and B22 bayonet lampholders must comply with BS EN 61184, and Regulation 559.6.1.7 requires a T2 temperature rating. This can easily be overlooked by designers and installers. Edison screw lampholders should comply with BS EN 60238:1999, which details temperature ratings.

555.1 The installation of autotransformers and step-up transformers must comply with the requirements of BS 7671. Particular care is needed to guard against danger if there is no neutral feed e.g. on primaries connected across the lines. Where there is a neutral it must be connected to the common terminal of the transformer. Step-up autotransformers are not permitted on an IT system. Multi-pole linked switches must be used so that all live conductors, including the neutral if any, can be simultaneously disconnected from the supply (see also Guidance Note 2).

8.5 Luminaires

8.5.1 Scope 559

Section 559 applies to the selection and erection of luminaires and lighting installations intended to be part of the fixed installation, and to highway power supplies and street furniture.

Particular requirements are given for:

1 fixed outdoor lighting installations
2 extra-low voltage lighting installations supplied from a source with a maximum rated voltage of 50 V a.c. or 120 V d.c.
3 lighting for display stands (See also requirements of Exhibitions, Shows and Stands, Section 711 of BS 7671 and Guidance Note 7)

Section 559 does not apply to:

4 high voltage signs supplied at low voltage (such as neon tubes)
5 signs and luminous discharge tube installations operating from a no-load rated output voltage exceeding 1 kV but not exceeding 10 kV (BS EN 50107). The requirements for high voltage signs are given in BS 559 and the BS EN 50107 series.

8.5.2 General

Luminaires should be manufactured and tested to the relevant standard, i.e BS EN 559.4.1
60598. Unfortunately, some cheaper or imported types may not be in compliance with BS EN 60598. The installer should be wary of the construction and heat dissipation of these products, and all luminaires should be installed strictly in accordance with the manufacturer's instructions.

Luminaires installed without transformers or convertors but fitted with extra-low 559.4.2
voltage lamps connected in series should be considered as low voltage equipment.

8.5.3 Protection against fire

Luminaires are usually installed on non-combustible surfaces, defined in BS EN 60598-1 as 'material incapable of supporting combustion' and taken to include metal, plaster and concrete. However, sometimes it may be necessary to mount luminaires on materials that may be considered to be flammable (e.g. wood and wood-based materials more than 2 mm thick) and in such cases a classified luminaire to BS EN 60598 marked with the following symbol should be selected.

Note: The luminaire standard BS EN 60598-1 is presently under review, with the 7th Edition anticipated to cease use of the present 'F' marking principles. Suitability of use for installation on flammable material will be presumed unless the luminaire is provided with information to the contrary.

Where luminaires are marked with this symbol, the excessive temperatures that may arise in operation, or due to the failure of a component, will not cause the mounting surface temperature to exceed 90 °C and thereby cause a risk of fire. Originally this mark only covered luminaires with integral controlgear but it now applies to all luminaires complying with the BS EN 60598 series of standards.

Luminaires which cannot comply with this criterion are required, at the time of writing (but see the above note), to carry a warning notice or have the following symbol:

It may be necessary to mount luminaires in areas where combustible dust could be present. In such cases a classified luminaire to BS EN 60598-2-24:1999 marked with the following symbols should be selected:

 and 

For simple pendants and wall lights the installation of the luminaire can have some effect on the compliance and it may be necessary for the designer to specify the minimum length of flexible cord or specific mounting details to limit the heat output on to local surfaces. This is especially significant for high power uplighters. (See Guidance Note 4 for further information on thermal effects and protection.)

For recessed luminaires, such as downlighters and ELV luminaires to BS EN 60598, the marking is amended to indicate the temperature rise criteria when in insulated ceilings or enclosed in thermal insulation, shown here:

Note: The 7th Edition of BS EN 60598-1 is anticipated to replace this symbol, and the one shown below, with others more descriptive of the situation.

For recessed luminaires not suitable for such installation the mark will be shown crossed out:

527.1.1 The fire rating of any ceiling or wall which is penetrated to install recessed luminaires must be maintained after the installation of the luminaires and Regulation 527.1.1 requires the designer to specify how this is to be carried out. Further information on fire protection requirements can be found in Approved Document B of the Building Regulations 2000 and Guidance Note 4. The luminaire manufacturer's advice and instructions must also be followed.

Similarly, Approved Documents C, *Site preparation and resistance to contaminants and moisture* and E, *Resistance to the passage of sound*, require maintenance of airtightness and acoustic ratings respectively and appropriately, following penetration of airtightness and acoustic barriers.

From a thermal point of view the temperature of illuminated objects is also important and the heat from a wrongly placed luminaire can initiate combustion. Luminaires to BS EN 60598, where required to be so, can also be marked with a symbol to indicate the minimum distance from a lighted object and a maximum lamp wattage.

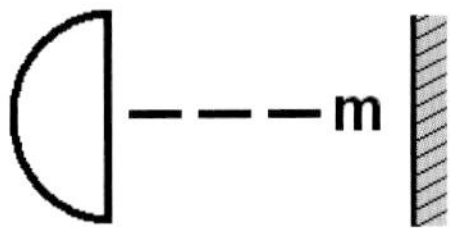

The lamp type is also important and dichroic lamps, where the heat is conducted away through the lamp and luminaire body rather than radiated as part of the light energy, must only be used in the correct types of luminaire. An alternative type of lamp, known as aluminised, is designed to emit both light and heat away from the luminaire. In doing so, attention is drawn especially to the minimum distance symbol shown above in relation to lighted objects and materials.

The luminaire manufacturer's instructions must be followed in all cases.

Regulations 422.3.1 and 422.4.2 give the following minimum distances at which 422.3.1
spotlights and projectors must be installed from combustible materials: 422.4.2

1 Rating up to 100 W – 0.5 m
2 Rating over 100 W up to 300 W – 0.8 m
3 Rating over 300 W up to 500 W – 1.0 m.

Section 526 requires joints between conductors or conductors and equipment to be
properly made and enclosed, and Regulation 526.5 details suitable enclosures. This 526.5
requirement applies to all types of connections in electrical circuits – even to 6 volt and
12 volt ELV luminaires. Regulation 526.9 requires the enclosure of the cores of cables 526.9
from which the sheath has been removed. Not only must the connection between the fixed wiring and the luminaire flexible cable be enclosed but the cores from which the sheath has been removed must also be enclosed. Exposed unsheathed flexible cable is not acceptable.

8.5.4 Transformers, convertors and controlgear

Where the SELV source of a SELV lighting system is a safety isolating transformer, it 559.11.3.1
should meet the requirements of BS EN 61558-2-6. Regulation 559.11.3.1 requires that the transformer has a protective device on the primary side or it should be short-circuit proof and marked with the following symbol:

Where an electronic convertor, such as a switch-mode power supply, is used to supply 559.11.3.2
an extra-low voltage lighting installation, it should comply with BS EN 61347-2-2 and be marked with the following symbol:

8.5.5 Independent lamp controlgear

559.7 Where lamp controlgear is suitable for mounting external to the luminaire and is mounted on a flammable surface, it must comply with one of the following requirements:

1 'class P' thermally protected ballast(s)/transformer(s), marked with the symbol:

2 temperature declared thermally protected ballast(s)/transformer(s), marked with the symbol:

with a marked value equal to or below 130 °C.

Note: The generally recognised symbol is of an independent ballast of BS EN 60417:

8.5.6 Stroboscopic effect

559.9 Stroboscopic effects can be extremely dangerous in particular circumstances. In the case of lighting for premises where machines with moving parts are in operation, proper consideration must be given to stroboscopic effects which can give a misleading impression of moving parts being stationary. This effect can be eliminated by using either luminaires provided with high frequency electronic controlgear or conventionally controlled luminaires installed with lagging and leading power-factor correction capacitors.

8.6 Highway power supplies, street furniture and external lighting installations

8.6.1 Protection against electric shock

Equipment used for highway power supplies is usually located in areas accessible to
559.10.1 the public. In view of this, using the measures of obstacles and placing out of reach to provide basic protection is not considered safe, i.e. these should be applied only in a controlled environment accessible only to skilled or instructed persons.

See 8.6.2 for the exceptions for overhead lines.

559.10.1

8.6.2 Overhead lines

Where low voltage overhead lines are constructed to a suitable standard complying with the Electricity Safety, Quality and Continuity Regulations 2002 and maintenance of nearby electrical equipment is by specially trained skilled persons only, the use of basic protection by placing out of reach is permissible.

Where such equipment is located within 1.5 m of an uninsulated overhead line, the protective measure of placing out of reach must not be used.

8.6.3 Equipment doors

Doors providing access to electrical equipment contained in street furniture provide a measure of protection against interference; however, the likelihood of removal or breakage is such that a door less than 2.5 m above ground level cannot be relied upon to provide basic protection. Equipment or barriers within the street furniture are required to prevent contact with live parts by a finger (IP2X or IPXXB). 559.10.3.1

A street furniture door which has no electrical equipment mounted upon it and is not likely to be contacted by wiring is neither an exposed-conductive-part nor an extraneous-conductive-part. Whilst the street lighting column or furniture frame may be an exposed-conductive-part, the door is not and therefore does not need to be earthed.

8.6.4 Distribution circuits

Items of fixed equipment supplied from a highway distribution circuit are generally in contact with the ground and they are no more likely to be subject to bodily contact during a fault condition than items of fixed equipment within a building. Regulation 559.10.3.3 of BS 7671 recognises a general 5 s maximum disconnection time for fixed equipment used in highway power supplies. 559.10.3.3

8.6.5 Devices for isolation and switching

Each item of street located equipment and street furniture is required to have a local means of isolation. The established practice of using the fuse carrier as the isolation and switching device is allowed for TN systems provided only instructed persons carry out the work. Formal instruction with the issue of authorization is appropriate, particularly if the electricity distributor's consent is required. Reference in Regulation 559.10.6.1 to a suitably rated fuse carrier will generally limit the use of the fuse carrier as the main switch to individual items of equipment, e.g. one street light, as the fuse carrier should generally not be used to switch currents exceeding 16 A. 559.10.6.1 559.10.6.2

8.6.6 Protection and identification of cables

The definitions in Regulation 1(5) of the ESQC Regulations are intended to include cables supplying street furniture within the scope of the statutory instrument. Such cables are included within the term 'network' and the persons owning or operating such cables are 'distributors' as defined. The implications of this are as follows:

- In accordance with Regulation 14, the cables must be buried at sufficient depth to prevent danger and they must be protected or marked.
- In accordance with Regulation 15(2), the distributor must maintain maps of the cables.

The DTI guidance on the ESQC Regulations gives advice on the methods that duty holders should employ to demonstrate compliance with Regulation 14(3) and thereby reduce the risk of injury to contractors or members of the public. Listed in order of preference the methods are:

1. cable installed in a duct with marker tape above
2. cable installed in a duct only
3. cable laid direct and covered with protective tiles
4. cable laid direct and covered with marker tape
5. some other method of mark or indication.

In consideration of the methods by which cables should be marked or protected, duty holders should make allowance for the environment in which the cables are installed and the risks to those who may need to expose and work on or near the cables in future.

The National Joint Utilities Group (NJUG) have agreed colours for ducts, pipes, cables and marker/warning tapes when laid in the public highway. The agreement for highway authority services is reproduced in Table 8.1. Highway distribution circuits mainly consist of cables buried underground. It is impracticable to provide permanent markings to indicate the position of all underground cables but records or drawings indicating the installed locations should be provided.

Note: These guidelines describe utility industry practice. However, it should not be assumed that all apparatus will conform to the recommendations for positioning and colour coding contained in this publication.

▼ **Table 8.1** Recommended colour coding of other underground apparatus (at the time of publication the following were current examples of known highway authority apparatus colour coding)

					Recommended minimum depths (mm)*	
Asset owner	**Duct**	**Pipe**	**Cable**	**Marker system**	**Footway**	**Carriageway**
			Street lighting			
England and Wales	Black or orange *Consult electricity company first*	N/A	Black	Yellow with black legend	450	600
Scotland	Purple	N/A	Purple	Yellow with black legend or purple	450	450
Northern Ireland	Orange	N/A	Black or orange	Various	450	450
			Other			
Traffic control	Orange		Orange	Yellow with black legend		
Street furniture	Black	N/A	Black	Yellow with black legend	450	600
Telecoms	Purple/orange	N/A	Black	Various		
			Motorways and trunk roads – England and Wales			
Communications	Purple	N/A	Grey	Yellow with black legend	450	
Communications power	Purple	N/A	Black	Yellow with black legend		
Road lighting	Orange	N/A	Black	Yellow with black legend		
			Motorways and trunk roads – Scotland			
Communications	Black or grey	N/A	Black	Yellow with black legend		
Road lighting	Purple	N/A	Purple	Yellow with black legend		

* All depths are from the surface level to the crown of the apparatus.

All cables buried direct are required to have a marker tape placed above them at a depth of 150 mm below finished ground level. Additional protection such as ducts or cable tiles would provide the degree of protection required by Regulation 522.8.10 where space is limited. Normally, cables buried direct 450 mm below verges or 750 mm below carriageways will have adequate physical protection. 522.8.10

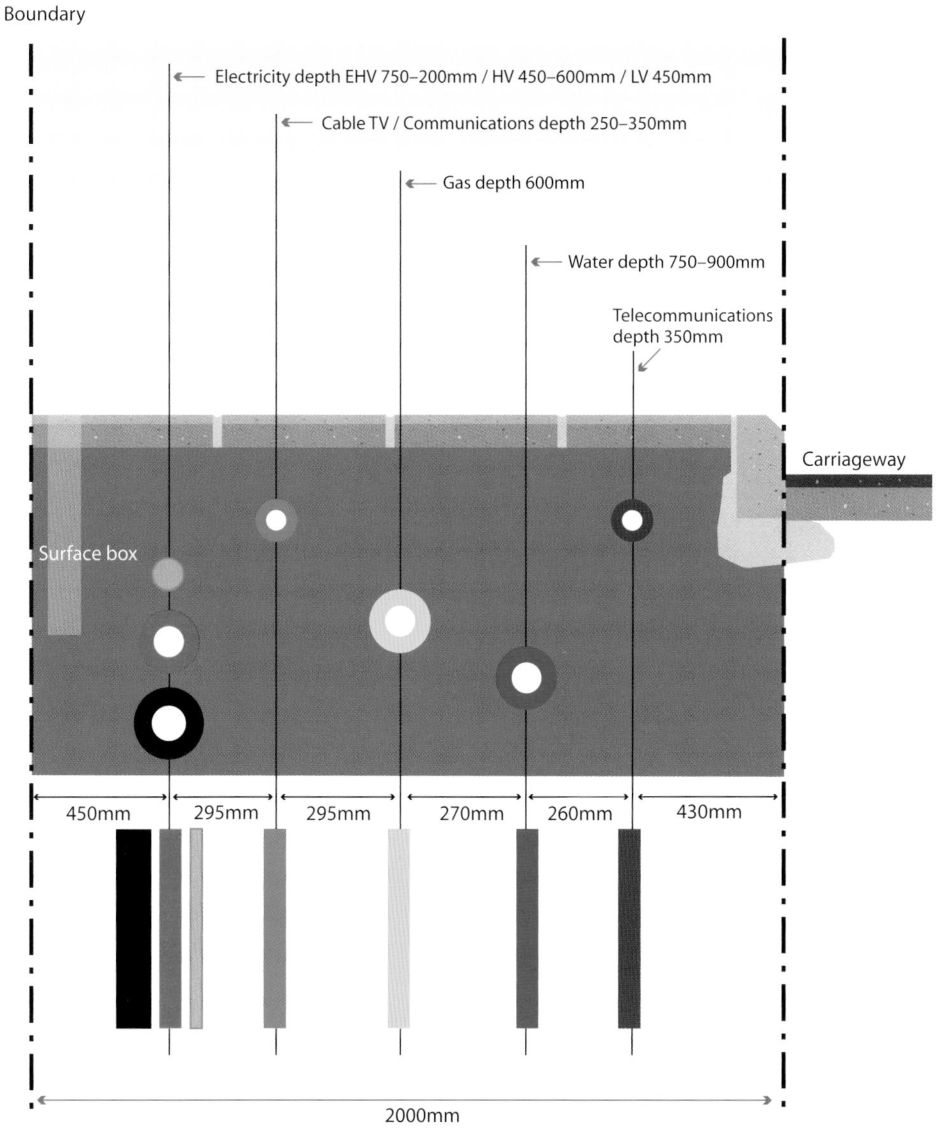

▼ **Figure 8.6** Recommended positioning of utility apparatus in 2 metre footway

8.6.7 External influences

512.2.1

Highway equipment suffers from many external influences, including condensation, corrosion, vibration, vandalism, etc., and the designer should take account of such factors in the selection of equipment. Where lamp controlgear is present, the heat generated by the equipment coupled with adequate ventilation will generally be sufficient to combat the effect of condensation and corrosion. However, in other situations the provision of a low wattage heater may be necessary to limit the effects of condensation and extend useful life. Luminaires and other electrical equipment should be secured in a manner to withstand vibration caused by wind and vehicles. All enclosures should be provided with access doors that can only be opened with a tool and are resistant to vandalism.

559.10.5.2 Equipment is required to have a degree of protection against ingress of solid objects and water after erection of at least IP33.

559.10.3.1 (v) It is necessary to reduce risks associated with imported and exported faults. Equipotential bonding to adjacent metallic structures could introduce a hazard during fault conditions by transmitting dangerous voltages above earth potential. Therefore, adjacent metallic structures should not be bonded to the circuit protective conductor or main earthing terminal without careful consideration of the possible consequences of doing so.

8.6.8 Temporary supplies

131.8 Before connecting any temporary load, the requirements which deal with increase of load must be considered.

It is important to protect against damage to the existing permanent equipment in the base compartment of an item of street furniture when connecting up a temporary installation. Where it is envisaged that temporary connections will be reused in the future, it is considered preferable to fit a socket-outlet to feed the temporary load, which can then be left in position after the load has been disconnected and reused at
559.10.7.2 a later date. An externally mounted label must be fixed stating the maximum sustained current which can be supplied from the outlet.

Also, it is intended that access to the base compartment of street furniture and street located equipment by unauthorized persons must not be made significantly easier by the addition of a temporary supply unit. Whatever the type of unit used, it must be capable of being secured in place by, for example, stainless steel banding and must totally cover the base compartment aperture. It is recommended that temporary supply units, which are designed to fit externally over the base compartment aperture of street furniture, are manufactured from corrosion-resistant material and fitted with suitable seals to prevent ingress of water.

8.6.9 Temporary decorative lighting

It is recommended that any decorative lighting (such as Christmas tree lights) within reach of the general public be supplied by SELV. Such equipment is readily available and suitable for indoor and outdoor use. Decorative lighting installed out of reach of the public may be supplied at 230 V with RCD protection at 30 mA. The risk of unwanted tripping may be minimized by splitting circuits.

559.10.3.2

8.6.10 Automatic disconnection of supply

BS 7671 carries a recommendation that equipment such as lighting arrangements for telephone kiosks, bus shelters, town plans, etc., is provided with additional protection by an RCD rated at 30 mA.

8.6.11 Class II equipment

Some items of street furniture such as some parking meters and some telephone
559.10.4 kiosks are of Class II construction. Regulation 559.10.4 requires that where the protective measure of double or reinforced insulation is used, any metal enclosure
412.2.2.4 of such equipment, such as the telephone kiosk structure, is not connected to the earthing terminal so that any problems associated with open-circuit PEN conductors
412.1.3 of PME supplies are avoided. However, the installation must then be under effective supervision in normal use to prevent changes to the installation that would impair the effectiveness of the protective measure.

8.6.12 References

- BS 5489 *Code of Practice for Road Lighting.*
- NJUG Publication 7 *Recommended positioning of utilities mains and plant for new works.*
- Electricity Association Engineering Recommendation G39/1 *Model code of practice covering electrical safety in the planning, installation, commissioning and maintenance of public lighting and other street furniture.*
- Institution of Lighting Engineers *Code of Practice for Electrical Safety in Public Lighting Operations.*

8.7 Extra-low voltage lighting installations

8.7.1 Protection against overcurrent

SELV circuits must be protected against overcurrent either by a common protective device or by a protective device for each SELV circuit, in accordance with the requirements of Chapter 43. When selecting the protective device for the primary circuit, account should be taken of the magnetising current of the transformer. The overcurrent protective device must be of the non-self-resetting type.

Chapter 43

8.7.2 Isolation and switching

Where transformers are operated in parallel, the primary circuit must be permanently connected to a common isolating device (see Figure 8.7). The transformers must also have identical electrical characteristics.

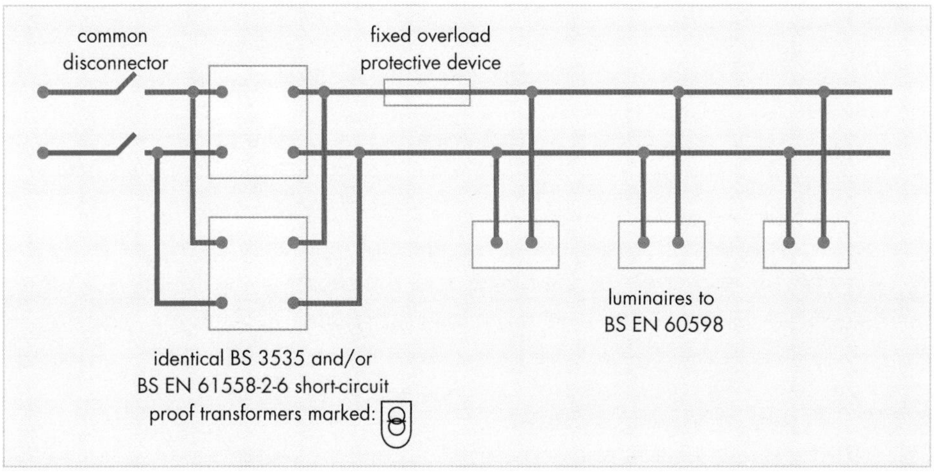

▼ **Figure 8.7**
SELV lighting installation with parallel operation of transformers and insulated conductors

8.7.3 Protection against fire

Extra-low voltage lamps and their transformers generate significant heat; fire risks arise from the installation of lamps close to floorboards or joists and the installation of transformers in high ambient temperatures and with limited ventilation. Also, the installer needs to ensure that thermal insulation materials do not interfere with the heat dissipation from the equipment.

8.7.4 Nature of processed or stored materials

The manufacturer's installation instructions must be followed, particularly those relating to mounting on flammable surfaces.

8.7.5 Fire risk of transformers/convertors

559.11.3.1 Transformers should be either:

- protected on the primary side by the protective device required for protection against fire, or
- short-circuit proof (both inherently and non-inherently proof).

559.11.3.2 Electronic convertors should comply with BS EN 61046 and with the requirements of BS EN 60598-2-23, clause 23.7.6.

Note: It is recommended that convertors marked with the following symbol are used:

8.7.6 Fire risk from short-circuiting of uninsulated conductors

559.11.4.1 If both circuit conductors are uninsulated, they must be either:

- provided with a special protective device complying with the requirements below, or
- the system must comply with BS EN 60598-2-23 (Particular requirements. ELV lighting systems for filament lamps).

559.11.4.2 Where the first option is adopted to provide protection against the risk of fire, the device must:

- continuously monitor the power demand of the luminaires;
- automatically disconnect the supply circuit within 0.3 s in the event of a short-circuit or failure that causes a power increase of more than 60 W;
- automatically disconnect while the supply circuit is operating with reduced power (for example, by gating control or a regulating process or a lamp failure) or if there is a failure that causes a power increase of more than 60 W;
- automatically disconnect during switching of the supply circuit if there is a failure that causes a power increase of more than 60 W; and
- be fail-safe.

8.7.7 Types of wiring system

The following wiring systems should be used:

- insulated conductors in conduit to BS EN 61386-21 or trunking or ducting to BS EN 50085-1;
- cables;
- flexible cables or cords;
- systems for extra-low voltage lighting according to BS EN 60598-2-23;
- track systems according to BS EN 60570.

Where parts of the extra-low voltage lighting installation are accessible, the requirements Section 423 of Section 423 of BS 7671 apply.

559.11.5.1 Metallic structural parts of buildings, e.g. pipe systems or parts of furniture, must not be used as live conductors.

Under no circumstances should conductors be used for other purposes, e.g. for supporting sign plates.

8.7.8 Bare conductors

559.11.5.3

If the nominal voltage is less than 25 V a.c. or 60 V d.c., bare conductors may be used provided that the extra-low voltage lighting installation complies with the following requirements:

- The lighting installation is designed and installed or enclosed in such a way that the risk of a short-circuit is reduced to a minimum, and
- the conductors used have a cross-sectional area of at least 4 mm^2, for mechanical reasons, and
- the conductors or wires are not placed directly on combustible material.

For suspended bare conductors, at least one conductor and its terminals must be insulated for that part of the circuit between the transformer and the protective device, to prevent the occurrence of short-circuit.

8.7.9 Suspended systems

559.11.6

If the fixing accessory is intended to support a pendant luminaire, the accessory should be capable of carrying a mass of not less than 5 kg. If the mass of the luminaire is greater than 5 kg, the installer should ensure that the fixing means is capable of supporting its mass. The installation instructions of the manufacturer should be followed.

Termination and connection of conductors should be by screw terminals or screwless clamping devices complying with BS EN 60998-2-1 or BS EN 60998-2-2.

Insulation piercing connectors and termination wires that rely on counterweights hung over suspended conductors to maintain the electrical connection must not be used.

The suspended system should be fixed to walls or ceilings by insulated distance cleats and must be continuously accessible throughout the route.

8.7.10 Cross-sectional area of conductors

559.11.5.2

The minimum cross-sectional area of the extra-low voltage conductors should be:

- 1.5 mm^2 copper, but in the case of flexible cables with a maximum length of 3 m a cross-sectional area of 1 mm^2 copper may be used;
- 4 mm^2 copper in the case of suspended flexible cables or insulated conductors, for mechanical reasons.

8.7.11 Electrical connections

All connections including ELV connections are required to meet the enclosure requirements of Section 526. This provides some protection against overheating in the event of a poor connection. 526.5

8.8 Selection and erection in relation to operation and maintainability

The Electricity at Work Regulations 1989 specifically place a duty on installation or equipment owners or managers (Duty Holders) to ensure the installed equipment is utilized safely and persons are protected from danger. In the workplace, ordinary persons are not permitted to access live systems and equipment or carry out electrical work.

All electrical installations should be as simple as possible. All manual components generally accessible should be of the simplest kind. Where not immediately obvious, operating instructions should be displayed adjacent to the component. All other equipment should be located behind lockable doors or accessible only by the use of a tool. The building operator should be provided with full instructions as to the operation and maintenance of the installation, including mechanical operation.

341.1 BS 7671 requires an assessment of expected maintenance to be made as part of the design process. Selection of equipment should be relevant to the maintenance (or lack of) it may receive during its installed and operating life.

For all items that provide physical protection which may need removal (including lids of cable boxes, trunking, etc.) and replacement, the procedure should be as simple as possible. It should be possible to be carried out by one person without assistance. Where frequent access is likely, lids, etc., should be hinged. It should also be possible to close all lids and covers without putting pressure on enclosed cables and equipment.

132.12 Electrical distribution equipment and switchgear should be located so that all
513.1 components may be operated and maintained safely. The Electricity at Work Regulations
529.3 1989 (Regulation 15) specifically require adequate working space, access and lighting to be provided for all electrical equipment, where working on or near it may give rise to danger.

559.6.2.1 In some instances it may be necessary to link the wiring through a luminaire on to the next luminaire. This method of installation should only be used if the luminaire(s) is designed for such practice.

559.6.2.2 A cable for through wiring must be selected in accordance with the temperature information on the luminaire or on the manufacturer's instruction sheet, if any, as follows:

1 For a luminaire complying with BS EN 60598 and with temperature marking, cables suitable for the marked temperature must be used

2 Unless specified in the manufacturer's instructions, for a luminaire complying with BS EN 60598 but with no temperature marking, heat-resistant cables are not required

3 In the absence of information, heat-resistant cables and/or insulated conductors of type H05S-U, H05S-K, H05SJ-K, H05SS-K (BS 6007) or equivalent must be used.

559.6.2.3 When groups of luminaires are divided across the three line conductors of a three-phase system with only one common neutral conductor, at least one device that simultaneously disconnects all line conductors should be installed.

132.12 Luminaires should be located so that they may be safely re-lamped and repaired
513.1 by maintenance personnel from a horizontal surface local to the luminaire. The maintenance personnel should be able to stand directly upon the surface or upon steps at a reasonable height. Where this cannot be achieved, consideration should be given to supplementary means of access and/or the luminaire should be equipped with hoisting access or hoisting equipment.

Cable capacities of conduit and trunking

A

A.1 General

This appendix describes a method based on practical work and experimentation which can be used to determine the size of conduit or trunking necessary to accommodate cables of the same size or differing sizes, and provides a means of compliance with the requirements of Chapter 52 of BS 7671.

The method employs a 'unit system', each cable size being allocated a factor. The sum of all factors for the cables intended to be run in the same enclosure is compared against the factors given for conduit, ducting or trunking, as appropriate, in order to determine the size of the conduit or trunking necessary to accommodate those cables.

It has been found necessary, for conduit, to distinguish between:

- case 1 – straight runs not exceeding 3 m in length, and
- case 2 – straight runs exceeding 3 m, or runs of any length incorporating bends or sets.

The term 'bend' signifies a British Standard 90° bend and one double set is equivalent to one bend.

For case 1, each conduit size is represented by only one factor. For case 2, each conduit size has a variable factor which is dependent on the length of run and the number of bends or sets. For a particular size of cable the factor allocated to it for case 1 is not the same as for case 2.

For trunking, each size of cable has been allocated a factor, as has been each size of trunking.

A number of variables affect any attempt to arrive at a standard method of assessing the capacity of conduit or trunking. Some of these are:

1. reasonable care in installation
2. acceptable use of the space available
3. tolerance in cable sizes
4. tolerance in conduit and trunking.

The following tables can only give guidance on the maximum number of cables which should be drawn in. The sizes should ensure an easy pull with low risk of damage to the cables.

Only the ease of drawing-in is taken into account. The electrical effects of grouping are not. As the number of circuits increases, the current-carrying capacity of the cables decreases. Thus cable sizes have to be increased, with the consequent increase in cost of cable and conduit.

It may sometimes be more attractive economically to divide the circuits concerned between two or more enclosures.

For sizing conduit and trunking the following three cases are dealt with:

Single-core thermoplastic (PVC) insulated cables to BS 6004 or single-core thermosetting cables to BS 7211:

1 in straight runs of conduit not exceeding 3 m in length (Tables A.1 and A.2)
2 in straight runs of conduit exceeding 3 m in length, or in runs of any length incorporating bends or sets (Tables A.3 and A.4)
3 in trunking (Tables A.5 and A.6).

Other sizes and types of cable in conduit or trunking are dealt with in Section A.5 of this appendix.

For cables and/or conduits not covered by this appendix, advice on the number of cables which can be drawn in should be obtained from the manufacturer.

A.2 Single-core thermoplastic (PVC) insulated cables in straight runs of conduit not exceeding 3 m in length

For each cable it is intended to use, obtain the appropriate factor from Table A.1.

Add the cable factors together and compare the total with the conduit factors given in Table A.2.

The minimum conduit size is that having a factor equal to or greater than the sum of the cable factors.

▼ **Table A.1**
Cable factors for use in conduit in short straight runs

Type of conductor	Conductor cross-sectional area (mm²)	Cable factor
Solid	1	22
	1.5	27
	2.5	39
Stranded	1.5	31
	2.5	43
	4	58
	6	88
	10	146
	16	202
	25	385

▼ **Table A.2**
Conduit factors for use in short straight runs

Conduit diameter (mm)	Conduit factor
16	290
20	460
25	800
32	1400
38	1900
50	3500
63	5600

A.3 Single-core thermoplastic (PVC) insulated cables in straight runs of conduit exceeding 3 m in length, or in runs of any length incorporating bends or sets

For each cable it is intended to use, obtain the appropriate factor from Table A.3.

Add the cable factors together and compare the total with the conduit factors given in Table A.4, taking into account the length of run it is intended to use and the number of bends and sets in that run.

The minimum conduit size is that size having a factor equal to or greater than the sum of the cable factors. For the larger sizes of conduit, multiplication factors are given relating them to 32 mm diameter conduit.

▼ **Table A.3**
Cable factors for use in conduit in long straight runs over 3 m, or runs of any length incorporating bends

Type of conductor	Conductor cross-sectional area (mm²)	Cable factor
Solid	1	16
or	1.5	22
stranded	2.5	30
	4	43
	6	58
	10	105
	16	145
	25	217

▼ Table A.4 Cable factors for runs incorporating bends and long straight runs

Length of run (m)	Conduit diameter (mm)																			
	16	20	25	32	16	20	25	32	16	20	25	32	16	20	25	32	16	20	25	32
	Straight				One bend				Two bends				Three bends				Four bends			
1	Covered by Tables A.1 and A.2				188	303	543	947	177	286	514	900	158	256	463	818	130	213	388	692
1.5					182	294	528	923	167	270	487	857	143	233	422	750	111	182	333	600
2					177	286	514	900	158	256	463	818	130	213	388	692	97	159	292	529
2.5					171	278	500	878	150	244	442	783	120	196	358	643	86	141	260	474
3					167	270	487	857	143	233	422	750	111	182	333	600				
3.5	179	290	521	911	162	263	475	837	136	222	404	720	103	169	311	563				
4	177	286	514	900	158	256	463	818	130	213	388	692	97	159	292	529				
4.5	174	282	507	889	154	250	452	800	125	204	373	667	91	149	275	500				
5	171	278	500	878	150	244	442	783	120	196	358	643	86	141	260	474				
6	167	270	487	857	143	233	422	750	111	182	333	600								
7	162	263	475	837	136	222	404	720	103	169	311	563								
8	158	256	463	818	130	213	388	692	97	159	292	529								
9	154	250	452	800	125	204	373	667	91	149	275	500								
10	150	244	442	783	120	196	358	643	86	141	260	474								

Additional factors:

- For 38 mm diameter use 1.4 x (32 mm factor)
- For 50 mm diameter use 2.6 x (32 mm factor)
- For 63 mm diameter use 4.2 x (32 mm factor)

A.4 Single-core thermoplastic (PVC) insulated cables in trunking

For each cable it is intended to use, obtain the appropriate factor from Table A.5.

Add the cable factors together and compare the total with the factors for trunking in Table A.6.

The minimum size of trunking is that size having a factor equal to or greater than the sum of the cable factors.

▼ **Table A.5** Cable factors for trunking

Type of conductor	Conductor cross-sectional area (mm^2)	Thermoplastic (PVC) BS 6004 Table 1 Cable factor	Thermosetting BS 7211 Table 3 Cable factor
Solid	1.5	8.0	8.6
	2.5	11.9	11.9
Stranded	1.5	8.6	9.6
	2.5	12.6	13.9
	4	16.6	18.1
	6	21.2	22.9
	10	35.3	36.3
	16	47.8	50.3
	25	73.9	75.4
	35	93.3	95.1
	50	128.7	132.8
	70	167.4	176.7
	95	229.7	227.0
	120	277.6	283.5
	150	343.1	346.4
	185	426.4	433.7
	240	555.7	551.6

Notes:

1. Cable factors are the cable cross-sectional area using the BS upper limit mean overall diameter
2. The provision of spare space is advisable; however, any circuits added at a later date must take into account grouping (see Appendix 4 of BS 7671 for further details)
3. Where thermosetting insulated conductors designed to operate at 90 °C (BS 5467 or BS 7211 etc.) are installed together with thermoplastic (PVC) insulated conductors designed to operate at 70 °C, it must be ascertained that the thermoplastic (PVC) insulated conductors will not be damaged. (Regulation 523.5.)

▼ **Table A.6**
Factors for trunking

Dimensions of trunking (mm x mm)	Trunking factor	Dimensions of trunking (mm x mm)	Trunking factor
50 x 38	767	200 x 100	8572
50 x 50	1037	200 x 150	13001
75 x 25	738	200 x 200	17429
75 x 38	1146	225 x 38	3474
75 x 50	1555	225 x 50	4671
75 x 75	2371	225 x 75	7167
100 x 25	993	225 x 100	9662
100 x 38	1542	225 x 150	14652
100 x 50	2091	225 x 200	19643
100 x 75	3189	225 x 225	22138
100 x 100	4252	300 x 38	4648
150 x 38	2999	300 x 50	6251
150 x 50	3091	300 x 75	9590
150 x 75	4743	300 x 100	12929
150 x 100	6394	300 x 150	19607
150 x 150	9697	300 x 200	26285
200 x 38	3082	300 x 225	29624
200 x 50	4145	300 x 300	39428
200 x 75	6359		

Note: These terms are for metal trunking with trunking thickness taken into account. They may be optimistic for plastic trunking, where the cross-sectional area available may be significantly reduced from the nominal by the thickness of the wall material.

A.5 Other sizes and types of cable in conduit or trunking, including flexible conduit

For sizes and types of cable in conduit or trunking other than those given in Tables A.1 to A.6, the number of cables installed should be such that the resulting space factor does not exceed 35% of the net internal cross-sectional area for conduit and 45% of the net internal cross-sectional area for trunking.

Flexible conduit types may have a smaller internal diameter due to increased wall thickness. The conduit manufacturer's advice should be obtained regarding cable capacity and cable grouping, and the required flexibility must be considered. The 35% space factor could also be utilized for flexible conduit.

Space factor is defined as the ratio (expressed as a percentage) of the sum of the overall cross-sectional areas of cables (insulation and any sheath) to the internal cross-sectional area of the conduit or other cable enclosure in which they are installed. The effective overall cross-sectional area of a non-circular cable is taken as that of a circle of diameter equal to the major axis of the cable.

The minimum internal radii of bends of cables for fixed wiring as given in Table G.1 should be used. Care should be taken to use bends in trunking systems, particularly with larger cables, that allow adequate bending radius.

A.6 Background to the tables

The 14th Edition of the *IEE Wiring Regulations* provided guidance on the number of cables which could be pulled into conduit. Unfortunately, the conduit capacities recommended could not always be achieved and guidance with regard to the effect of length, number of bends, etc., was only subjective. Further, the tables had been constructed with the use of an arbitrary space factor of 40% for conduit and 45% for trunking; the conduit factor was later shown to be inappropriate.

The criteria adopted for replacement capacity tables were that the length between pulling-in points should permit 'easy drawing in' and the cables should not be damaged. These were quite acceptable but, because damage and easy drawing-in were not defined, the selection of an appropriate size of conduit for a group of cables remained, to a large extent, dependent on the experience of the installer.

Practical pulling-in of cables was carried out over a wide range of cable and conduit arrangements to provide a rational means of predicting the size of conduit to accommodate a given bunch of mixed sizes of conduit cables (insulated conductors). The scope was limited to conduit sizes from 16 to 32 mm diameter in both steel and plastic and to single-core thermoplastic (PVC) insulated copper conductors from 1 to 25 mm^2. Single-wire conductors were included for the 1, 1.5 and 2.5 mm^2 sizes and multi-wire conductors were covered for sizes from 1.5 to 25 mm^2. Criteria were that the size of conduit should permit an 'easy pull-in' without insulation damage.

Empirical relationships between numbers and sizes of cables, distance between pull-in points, size of conduit and pulling force were deduced from numerous tests with straight conduit runs using a variety of makes of cable. The effects of type of conduit and ambient temperature were included in the investigation. Considerable variation in results was experienced but it was possible to determine design centre values for conduit capacities based on a 'unit' approach. The unit system included both solid and stranded conductor cables.

Simple mathematical models were developed during the analysis of the above results which expressed design centre values for conduit capacities using a unit system. For straight conduit runs of 3 m length and upwards and for all lengths which include bends or sets, these models can be used to produce conduit capacity tables or can be adapted for CAD applications. For lengths up to 3 m having no bends or sets, a separate empirical table of capacities was developed.

It should be noted that there is no interrelationship between any of the series of factors for the two different conduit installation cases or for trunking installation, the factors being developed separately for each system.

Degrees of protection provided by enclosures

B

B.1 IP code for ingress protection

B.1.1 General

The requirements of the IP code are given in BS EN 60529:2004. This is a standard used to form the basic requirements of electrical equipment standards and the construction for IP ratings of a specific type of equipment is given in the BS for that type of equipment.

The degree of protection provided by an enclosure is indicated by two numerals followed by an optional additional letter and/or optional supplementary letter(s) as shown in Figure B.1.

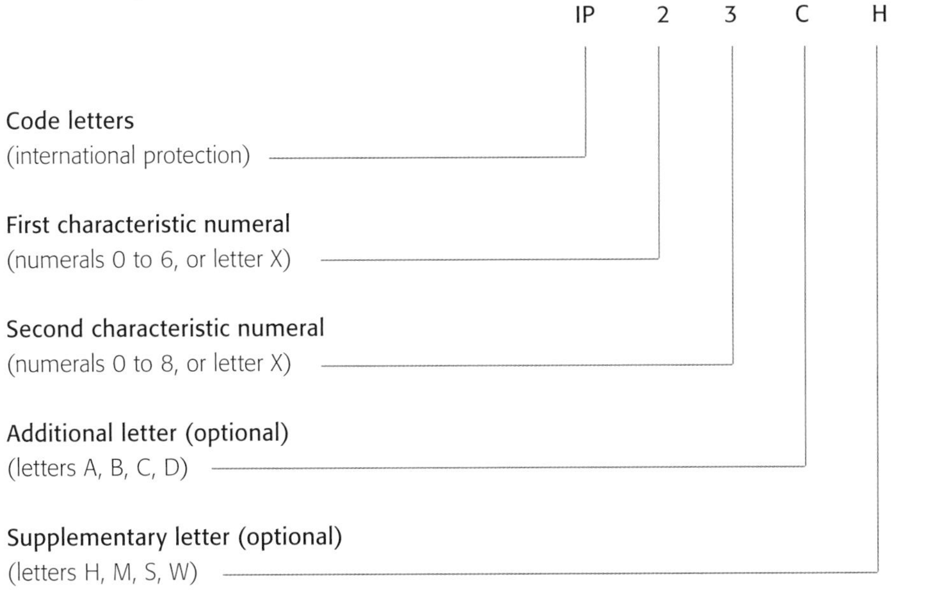

▼ **Figure B.1**
IP code format

Where a characteristic numeral is not required to be specified, it can be replaced by the letter 'X' ('XX' if both numbers are omitted).

The additional letter and/or supplementary letter(s) may be omitted without replacement. Where more than one supplementary letter is used, the alphabetic sequence is used.

Table B.1
IP code characteristic numerals

First characteristic numeral	Second characteristic numeral
(a) Protection of persons against access to hazardous parts inside enclosures (b) Protection of equipment against ingress of solid foreign objects	Protection of equipment against ingress of water

No.	Degree of protection	No.	Degree of protection
0	(a) Not protected (b) Not protected	0	Not protected
1	(a) Protection against access to hazardous parts with the back of the hand (b) Protection against foreign solid objects of 50 mm diameter and greater	1	Protection against vertically falling water drops
2	(a) Protection against access to hazardous parts with a finger (b) Protection against solid foreign objects of 12.5 mm diameter and greater	2	Protected against vertically falling water drops when enclosure tilted up to 15°. Vertically falling water drops shall have no harmful effect when the enclosure is tilted at any angle up to 15° from the vertical
3	(a) Protection against contact by tools, wires or such like more than 2.5 mm thick (b) Protection against solid foreign objects of 2.5 mm diameter and greater	3	Protected against water spraying at an angle up to 60° on either side of the vertical
4	(a) As 3 above but against contact with a wire or strips more than 1.0 mm thick (b) Protection against solid foreign objects of 1.0 mm diameter and greater	4	Protected against water splashing from any direction
5	(a) As 4 above (b) Dust-protected (dust may enter but not in amount sufficient to interfere with satisfactory operation or impair safety)	5	Protected against water jets from any direction
6	(a) As 4 above (b) Dust-tight (no ingress of dust)	6	Protected against powerful water jets from any direction
	No code	7	Protection against the effects of temporary immersion in water. Ingress of water in quantities causing harmful effects is not possible when enclosure is temporarily immersed in water under standardized conditions.
	No code	8	Protection against the effects of continuous immersion in water under conditions agreed with a manufacturer

B.1.2 Additional letter

The 'additional' letter indicates the degree of protection of persons against access to hazardous parts. It is only used if the protection provided against access to hazardous parts is higher than that indicated by the first characteristic numeral, or if only the protection against access to hazardous parts and not general ingress is indicated, the first characteristic numeral being then replaced by an X.

For example, such higher protection may be provided by barriers, suitable shape of openings or clearance distances inside the enclosure.

Refer to BS EN 60529 for full details of the tests and test devices used.

▼ **Table B.2** IP code additional letters

Additional letter	Brief description of protection
A	Protected against access with the back of the hand (minimum 50 mm diameter sphere) (adequate clearance from live parts)
B	Protected against access with a finger (minimum 12 mm diameter test finger, 80 mm long) (adequate clearance from live parts)
C	Protected against access with a tool (minimum 2.5 mm diameter tool, 100 mm long) (adequate clearance from live parts)
D	Protected against access with a wire (minimum 1 mm diameter wire, 100 mm long) (adequate clearance from live parts)

The classification of IPXXB, used frequently in BS 7671, indicates that there is no classification for water or dust ingress but that protection is provided against access to live parts with a finger. Similarly with IPXXD, also used in BS 7671, but where protection is provided against access with a wire.

B.1.3 Supplementary letter

In the relevant product standard, supplementary information may be indicated by a supplementary letter following the second characteristic numeral or the additional letter.

The following letters are currently in use but further letters may be introduced by future product standards.

▼ Figure B.3
IP code supplementary letters

Letter	Significance
H	High voltage apparatus
M	Tested for harmful effects due to the ingress of water when the movable parts of the equipment (e.g. the rotor of a rotating machine) are in motion
S	Tested for harmful effects due to the ingress of water when the movable parts of the equipment (e.g. the rotor of a rotating machine) are stationary
W	Suitable for use under specified weather conditions and provided with additional protective features or processes

B.1.4 Product marking

The requirements for marking a product are specified in the relevant product standard. An example could be:

IP23CS
An enclosure with this designation (IP code):

2 protects persons against access to hazardous parts with a finger and protects the equipment inside the enclosure against ingress of solid foreign objects having a diameter of 12.5 mm and greater.

3 protects the equipment inside the enclosure against the harmful effects of water sprayed against the enclosure at an angle of up to 60° either side of the vertical.

C protects persons handling tools having a diameter of 2.5 mm and greater and a length not exceeding 100 mm against access to hazardous parts (the tool may penetrate the enclosure up to its full length).

S is tested for protection against harmful effects due to the ingress of water when all the parts of the equipment are stationary.

If an enclosure provides different degrees of protection for different intended mounting arrangements, the relevant degrees of protection must be indicated by the manufacturer in the instructions relating to the respective mounting arrangements.

B.1.5 Drip proof and splash proof

Certain electrical equipment has been graded against water ingress by the identifications 'Drip Proof', etc. This labelling has been superseded by the IP coding but is given in Figure B.2 with an equivalent IP rating for comparison. Comparisons are not exact and the designer or installer must satisfy themselves that any equipment is suitable for the location of use.

Drip proof	IPX2
Rain proof	IPX3
Splash proof	IPX4
Jet proof	IPX5
Protected against immersion in water	IPX7

▼ **Figure B.2** Superseded water ingress labels and their equivalent IP ratings

B.2 IK code for impact protection

B.2.1 General

BS EN 62262:2002 *Degrees of protection provided by enclosures for electrical equipment against external mechanical impacts (IK code)* specifies a system for classifying the degrees of protection provided by enclosures against mechanical impact.

The Standard describes only the general requirements and designations for the system – the application of the system to a specific enclosure type will be covered by the British Standard applicable to that equipment or enclosure. An enclosure is defined as a part providing protection of equipment against certain external influences and protection against contact. This may be considered to include conduit, trunking, etc.

In general, the degree of protection will apply to a complete enclosure. If parts of an enclosure have different degrees of protection, they must be separately identified. The coding is separate from the IP rating and will be marked separately in the following way:

IK 05

Code letters
(international mechanical protection) — IK

Characteristic group numeral
(00 to 10) — 05

Each characteristic group numeral represents an impact energy value as shown below:

IK Code	IK00	IK01	IK02	IK03	IK04	IK05	IK06	IK07	IK08	IK09	IK10
Impact energy in joules	*	0.15	0.2	0.35	0.5	0.7	1	2	5	10	20

* No protection specified

When higher impact energy is required the value of 50 joules is recommended.

Standard circuit arrangements and the provision of socket-outlets

C

C.1 General

This appendix gives details of standard circuit arrangements that satisfy the requirements of Chapter 43 for overload protection and Section 537 for isolation and switching, together with the requirements as regards current-carrying capacities of conductors prescribed in Chapter 52 – Selection and erection of wiring systems and Appendix 4 of BS 7671.

It is the responsibility of the designer and installer when adopting these circuit arrangements to take the appropriate measures to comply with the requirements of other chapters or sections which are relevant, such as Chapter 41 'Protection against electric shock', Section 434 'Protection against fault current', Chapter 54 'Earthing and protective conductors', and the requirements of Chapter 52 other than those concerning current-carrying capacities.

Circuit arrangements other than those detailed in this appendix are not precluded where they meet the requirements of BS 7671.

The standard circuit arrangements are:

- final circuits using socket-outlets complying with BS 1363-2 and/or connection units complying with BS 1363-4
- radial final circuits using socket-outlets complying with BS EN 60309-2 (BS 4343)
- cooker final circuits in household or similar premises using cooker control units complying with BS 4177 or control switches complying with BS 3676 (dual numbered with BS EN 60669-1)
- water heater circuits
- electric showers.

C.2 Final circuits using socket-outlets and fused connection units complying with BS 1363

C.2.1 Layout

A ring, with spurs if any, or radial final circuit, feeds permanently connected equipment and an unlimited number of socket-outlets. Ring final circuits shall comply with the requirements of Regulation 433.1.5 of BS 7671 and the circuit conductors are to form a ring (with spurs as necessary) starting and finishing at the distribution board or consumer unit protective device, neutral bar and, where appropriate, earth bar. 433.1.5

The floor area served by the circuit is determined by the known or estimated load but does not exceed the value given in Table C.1.

311.1 An assessment of the loading must be made for the design of an installation, in accordance with Chapter 31 of BS 7671, and adequate circuits provided. (See also
132.3 Regulation 132.3.)

For household installations, a single 30 A or 32 A ring final circuit may serve a floor area of up to 100 m^2 but consideration should be given to the loading in kitchens which may require a separate circuit. Socket-outlets for washing machines, tumble driers and dishwashers should be located so as to provide reasonable sharing of the load in each leg of the ring, or consideration should be given to a 4 mm^2 radial circuit. 2.5 mm^2 radial circuits may also adequately serve areas of a dwelling other than kitchens. For other types of premises, final circuits complying with Table C.1 may be installed where, owing to diversity, the maximum demand of current-using equipment to be connected is estimated not to exceed the corresponding ratings of the circuit cables.

553.1.7 The number of socket-outlets should be such as to ensure compliance with Regulation 553.1.7, each socket-outlet of a twin or multiple socket-outlet unit being regarded as one socket-outlet.

Diversity between socket-outlets and permanently connected equipment has already been taken into account in Table C.1 and no further diversity may be applied.

▼ **Table C.1**
Typical final circuits using BS 1363 socket-outlets in household installations

Type of circuit		Overcurrent protective device rating (A)	Minimum conductor cross-sectional area* (mm^2)		Maximum floor area served‡ (m^2)
			Copper conductor thermoplastic or thermosetting insulated cables	Copper conductor mineral insulated cables	
1	2	3	4	5	6
A1	Ring	30 or 32	2.5	1.5†	100
A2	Radial	30 or 32	4	2.5	75
A3	Radial	20	2.5	1.5†	50

* The tabulated values of conductor size may need to be increased where more than two circuits are grouped together but may be reduced for fused spurs. Table C.1 applies to thermoplastic or thermosetting insulated cables of the types identified in Table 4A2 of Appendix 4 of BS 7671 in installation method C and method B and installed in accordance with those installation methods, and mineral insulated cable (MICC) as identified in Table 4G1A installed in accordance with reference methods C, E, F and G of Table 4A2. It does not apply to 70 °C thermoplastic (PVC) insulated and sheathed flat twin cable with protective conductor installed in accordance with method A of Table 4D5.

† 1.5 mm^2 two-core MICC installed for a ring final circuit with a protective device rated at not more than 32 A or a radial circuit with a protective device rated at not more than 20 A may be taken as having a conductor not operating at a temperature above 70 °C and may be used to comply with the relevant requirements of Regulation 512.1.2. 1.5 mm^2 four-core MICC will not comply with Regulation 433.1.5 at 70 °C.

‡ The socket-outlets are to be as evenly distributed between circuits as is reasonable.

Where two or more ring final circuits are installed, the socket-outlets and permanently connected equipment to be served should be reasonably distributed among the circuits.

C.2.2 Circuit protection

Table C.1 is applicable for circuits protected by fuses to BS 3036, BS 1361 and BS 88 and circuit-breakers, types B and C to BS EN 60898 or BS EN 61009-1 and BS EN 60947-2 and types 1, 2 and 3 to BS 3871.

C.2.3 Conductor size

The minimum size of conductor in the circuit and in non-fused spurs is given in Table C1. However, if the cables of more than two circuits are bunched together or the ambient temperature exceeds 30 °C, the size of conductor is increased and is determined by applying the appropriate rating factors from Appendix 4 of BS 7671 such that the size then corresponds to a current-carrying capacity not less than:

- 20 A for ring circuit A1 433.1.5
- 30 A or 32 A for radial circuit A2 (i.e. the rating of the overcurrent protective device)
- 20 A for radial circuit A3 (i.e. the rating of the overcurrent protective device).

The conductor size for a fused spur is determined from the total current demand served by that spur, which is limited to a maximum of 13 A.

Where a fused spur serves socket-outlets the minimum conductor size is:

- 1.5 mm^2 for thermoplastic (PVC) or thermosetting insulated cables, with copper conductors
- 1 mm^2 for mineral insulated cables, with copper conductors.

For a 30 A or 32 A ring final circuit with a protective device in accordance with the requirements of Regulation 433.1.5, supplying 13 A socket-outlets to BS 1363, a conductor with a minimum csa of 2.5 mm^2 is deemed to comply with the requirements of Regulation 433.1.1 if the current-carrying capacity (I_z) of the cable is not less than 20 A and if, under the intended conditions of use, the load current in any part of the ring is unlikely to exceed for long periods the current-carrying capacity (I_z) of the cable. 433.1.5 433.1.1

Electrical accessories to BS 1363, etc., are designed for a maximum operating temperature at their terminals and also for use with thermoplastic (PVC) insulated conductors operating at a maximum conductor temperature of 70 °C.

Thermosetting cables can operate with conductor temperatures up to 90 °C and some types of mineral insulated cable can operate with conductor temperatures up to 105 °C.

Such temperatures can damage accessories (see Regulation 512.1.2 of BS 7671). 512.1.2
Consequently, when used, such cables shall be derated to comply with this requirement (see notes 1 and 2 of Table 4E2A, etc. of Appendix 4 of BS 7671).

C.2.4 Spurs

The total number of fused spurs is unlimited but the number of non-fused spurs is not to exceed the total number of socket-outlets and items of stationary equipment connected directly in the circuit.

A non-fused spur feeds only one single or one twin or multiple socket-outlet or one permanently connected item of electrical equipment. Such a spur is connected to a circuit at the terminals of socket-outlets or at junction boxes or at the origin of the circuit in the distribution board.

A fused spur is connected to the circuit through a fused connection unit, the rating of the fuse in the unit not exceeding that of the cable forming the spur and, in any event, not exceeding 13 A. The number of socket-outlets which may be supplied by a fused spur is unlimited.

C.2.5 Permanently connected equipment

Permanently connected equipment, such as built-in washing machines, electric ovens, etc., is often connected to the supply by means of a plug and socket-outlet located
537.5.1.3 behind the appliance. A separate control switch need not be installed provided that:

- a functional switch is incorporated in the appliance, and
- the appliance is locally protected by a fuse to BS 1362, of a rating not exceeding 13 A, or
- the circuit is protected by a circuit-breaker of a rating not exceeding 16 A.

C.3 Radial final circuits using 16 A socket-outlets complying with BS EN 60309-2 (BS 4343)

C.3.1 General

Where a radial final circuit feeds equipment the maximum demand of which, having allowed for diversity, is known or estimated not to exceed the rating of the overcurrent protective device and, in any event, does not exceed 20 A, the number of socket-outlets is unlimited.

C.3.2 Circuit protection

The overcurrent protective device has a rating not exceeding 20 A.

C.3.3 Conductor size

The size of conductor is determined from Appendix 4 of BS 7671 by applying the appropriate rating factors and is such that it then corresponds to a current-carrying capacity not less than the rating of the overcurrent protective device.

C.3.4 Types of socket-outlet

Socket-outlets have a rated current of 16 A and are of the type appropriate to the number of phases, circuit voltage and earthing arrangement. Socket-outlets incorporating pilot contacts are not included.

C.4 Cooker final circuits in household premises

The final circuit supplies a control switch complying with BS 3676 or a cooker control unit (with a socket-outlet incorporated) complying with BS 4177. Assessment of the risks associated with the installation of a cooker control unit should always be made. A socket-outlet near a cooker could invite accidents if inappropriately located, by allowing cables or equipment near hot surfaces. A cooker control unit with a socket-outlet should be installed so as to minimize the risk of such accidents; if this is not possible it is perhaps better that cooker control units should not incorporate a socket-outlet. A modern kitchen should have an adequate number of socket-outlets.

The rating of a cooking appliance circuit is determined by the assessment of the current demand of the cooking appliance(s), and control unit socket-outlet if any, in accordance with Table H.1 of Appendix H. A 30/32 A circuit is suitable for most household cookers but a 40/45 A circuit may be necessary for larger cookers.

A circuit of rating exceeding 15 A but not exceeding 50 A may supply two or more cooking appliances where these are installed in one room. One switch may be used to control all the appliances. The control switch should be placed in a readily accessible 537.5.1.3
position. Attention is drawn to the need to afford discriminative operation of protective devices as stated in Regulation 536.1. 536.1

C.5 Water heating and electric shower final circuits in household premises

The final circuit shall supply the full load of the electric shower. No diversity is allowable on the final circuit.

Immersion heaters fitted to storage vessels in excess of 15 litres capacity, or permanently connected heating appliances forming part of a comprehensive electric space heating installation, should be supplied by their own separate circuits.

Immersion heaters should not be connected by a plug and socket-outlet but by a 554.3.3
switched connection unit complying with BS 1363-4.

C.6 Provision of socket-outlets

There is no statutory requirement for the minimum provision of socket-outlets but the number of socket-outlets needed in a dwelling has risen significantly over the years and can be expected to continue to rise as more electrical goods become available and people's working practices change, with working from home now becoming quite common. Table C.2 makes proposals for the provision of socket-outlets in modern dwellings.

553.1.7 C.6.1 Number of socket-outlets

To satisfy Regulation 553.1.7 sufficient socket-outlets must be provided and located to allow the use of mobile equipment with the length of flexible cord normally fitted. Long extension leads, trailing sockets and extending flexible cords which are incorrectly or inappropriately located can create tripping hazards and can damage and cause wear and overheating of the flex when flexible cords are placed under carpets, etc. In order, so far as is possible, to make the use of multiway adaptors or extension leads unnecessary, an adequate number of fixed socket-outlets should be installed as listed in Table C.2. Multi-way adaptors, long extension leads, trailing socket-outlets and extending flexible cords should never be used unattended and should always be visible and accessible throughout their length.

Unfortunately, no matter how good the overall layout some socket-outlets will inevitably become inaccessible behind furniture when the dwelling is furnished and the designer should consider this in the initial provisions. Also, the continued future growth in appliances and equipment should be expected and so what may seem initially an over-generous provision ultimately may not be.

▼ **Table C.2** Minimum number of twin socket-outlets to be provided in homes

Room type	Smaller rooms (up to 12 m²)	Medium rooms (12–25 m²)	Larger rooms (more than 25 m²)
Main living room (note 4)	4	6	8
Dining room	3	4	5
Single bedroom (note 3)	2	3	4
Double bedroom (note 3)	3	4	5
Bedsitting room (note 6)	4	5	6
Study	4	5	6
Utility room	3	4	5
Kitchen (note 1)	6	8	10
Garage (note 2)	2	3	4
Conservatory	3	4	5
Hallway	1	2	3
Loft	1	2	3
Locations containing a bath or shower		note 5	

(Table by kind permission of the ECA, Select and BEAMA.)

Notes to Table C.2:

1 **KITCHEN** – If a socket-outlet is provided in the cooker control unit, this should not be included in the 6 recommended in the table above.
Appliances built into kitchen furniture (integrated appliances) should be connected to a socket-outlet or switch fused connection unit that is accessible when the appliance is in place and in normal use. Alternatively, where an appliance is supplied from a socket-outlet or a connection unit, these should be controlled by an accessible double-pole switch or switched fused connection unit.
It is recommended that wall mounted socket-outlets above a work surface are spaced at not more than 1 m intervals along the surface.

2 **GARAGE** – The number of socket-outlets specified allows for the use of a battery charger, tools, portable light and garden appliances.

3 **BEDROOM** – It is envisaged that this room will be used in different ways in different households. It may be used simply as a child's bedroom requiring socket-outlets for table lamps, an electric blanket and an electric heater only; or it may serve as a teenager's bedroom and living room combined, where friends are entertained. In this case, socket-outlets may be needed for computers (printers, scanners, speakers,etc.), games consoles, MP3/4 players, mobile phone chargers, DVD players, digital receivers, home entertainment systems (amplifier, CD player), hairdryer, television and radio, in addition to lamps, an electric blanket and electric heater.

4 **HOME ENTERTAINMENT** – In addition to the number of socket-outlets shown in the table it is recommended that at least two further double socket-outlets are installed in home entertainment areas.

5 **LOCATIONS CONTAINING A BATH OR SHOWER** – Except for SELV socket-outlets 701.512.3
complying with Section 414 and shaver supply units complying with BS EN 61558-2-5, socket-outlets are prohibited within a distance of 3 m horizontally from the boundary of zone 1.

6 **BEDSITTING ROOM** – Rooms specifically designed or envisaged to be used as student bedsitting rooms should be provided with additional socket-outlets which may be needed since persons using these rooms will often introduce other portable appliances in addition to items already mentioned in Note 3. In such situations a lack of sufficient socket-outlets is an additional danger and therefore the minimum number of twin outlets should be increased to four.

C.6.2 Sinks and electrical accessories

How close are electrical accessories, such as socket-outlets, permitted to be to a sink? This question is commonly asked, not only by members of the public but also by electrical contractors and designers concerning both domestic and commercial premises.

Some are under the impression that a minimum distance is given in BS 7671 within which electrical accessories must not be installed. This is not true. It is impractical to prescribe such a distance in BS 7671, given the fact that many kitchens are very small. It is almost inevitable therefore, in many situations, that socket-outlets will be installed which are simultaneously accessible to sink tops.

The type of accessory installed must be of a design appropriate to its location as required by Regulations 132.5.1 and 512.2.1. Where BS 1363 accessories are used, 132.5.1
they should be installed as far from the sink as possible. As a general recommendation, 512.2.1
BS 1363 accessories should be mounted above sink height at a horizontal distance of not less than 300 mm from the edge of the sink top.

C.7 Mounting heights of accessories

553.1.6 Accessories must be mounted at a height that will make them accessible but minimize the risk of mechanical damage by their being struck by floor cleaning equipment or the movement of furniture, and allow a comfortable bending radius for larger or stiff flexible cords. The Building Regulations 2000 Approved Document M, Access and facilities for disabled people, contains appropriate mounting heights between 450 mm and 1200 mm for electrical equipment in dwellings (see Figure C.1).

▼ **Figure C.1**
Height of switches, sockets, etc. (see Approved Document M, section 8)

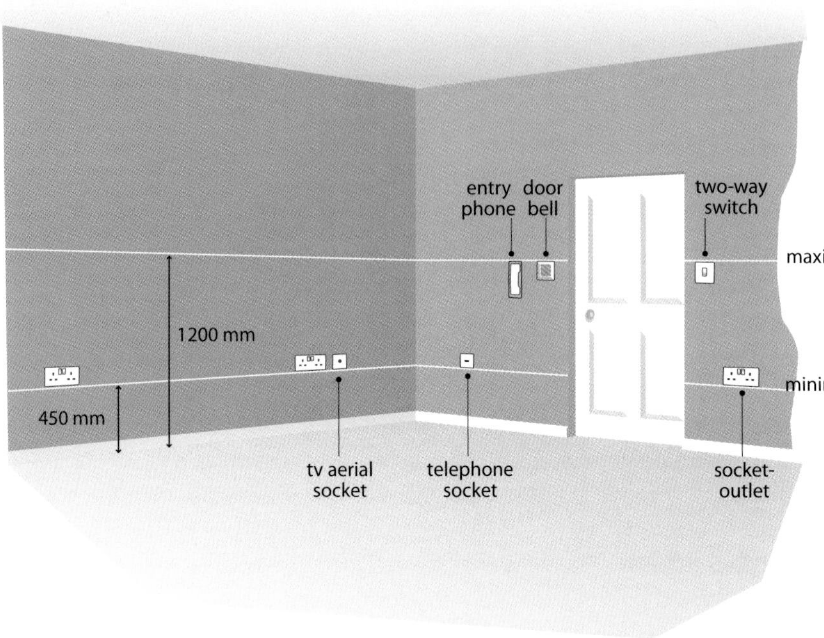

C.7.1 BS 8300:2001 *Design of buildings and their approaches to meet the needs of disabled people – Code of Practice*

BS 8300, which applies to all types of buildings, contains a number of recommendations which could affect the design of an electrical installation. The requirements concerning the location of outlets, switches, controls and meters are shown in the extract from the standard given in Figure C.2.

It should be noted that this document is not mandatory unless directly referred to in the specification or the specification contains a general requirement to the effect that the electrical installation shall comply with all relevant standards and codes. In case of doubt, the exact requirements of the client must be ascertained prior to undertaking the project.

▼ **Figure C.2** Extract from BS 8300:2001

10.5.2 Location of outlets, switches, controls and meters

All outlets, switches and controls, including two-way switching, should be positioned consistently in relation to doorways and corners within a building and in a logical sequence to suit passage through the building.

Note 1: By using vertical strips as switches, instead of single height switches, users can operate the switch at whichever height is convenient for them.

Preferably, light switches should align horizontally with door handles for ease of location when entering a room.

Electrical socket-outlets, telephone points and TV sockets should be located at least 400 mm but not more than 1000 mm above the floor. Socket-outlets whose plugs are likely to be removed and replaced frequently should be located at the top of the range (see figure below).

Note 2: Switches close to the floor or skirting are difficult and dangerous because they require users to stoop or kneel to operate them. The higher the socket-outlet, the easier it is to push in or pull out the plug.

Switches for permanently wired appliances (e.g. fused spurs or reset switches for alarm calls) should be mounted within the range between 750 mm and 1200 mm (see figure below).

Meters should be mounted between 1200 mm and 1400 mm from the floor so that the readings can be viewed by a person standing or sitting. Pre-pay meters should be accessible but protected so children cannot tamper with them.

All switches and controls that require precise hand movement/dexterity, e.g. for heating installations, ventilation, etc., should be in a zone 750 mm to 1000 mm from the floor (see figure below) so that wheelchair users and those standing can operate them.

The maximum height of simple push button controls, including isolator switches and circuit-breakers, that require limited dexterity should be 1200 mm (see figure below).

Outlets, switches and controls should be at least 350 mm from room corners.

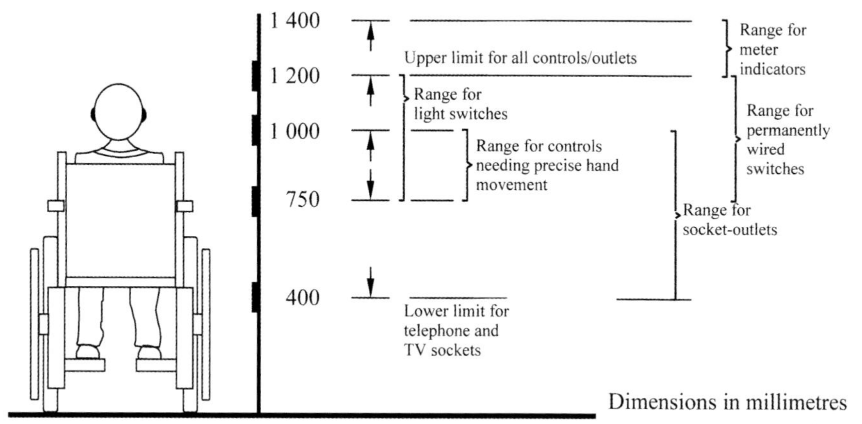

Limitation of earth fault loop impedance for compliance with Regulation 543.1.1

D

Regulation 543.1.1 indicates that the cross-sectional area of a protective conductor, other than a protective bonding conductor, shall be: 543.1.1

1 calculated in accordance with Regulation 543.1.3, or
2 determined in accordance with Regulation 543.1.4.

In some cases the type of wiring system selected for use will determine which of the two methods can be followed. For instance, the widely used flat twin and flat three-core thermoplastic (PVC) insulated and thermoplastic (PVC) sheathed cables with protective conductors (cables to Table 5 of BS 6004) do not comply with Table 54.7 of Regulation 543.1.4 (other than the 1 mm^2 size) and therefore Method 1 should be used.

Where Method 1 is used, in order to apply the formula given in Regulation 543.1.3 it is essential that the time/current characteristic of the overcurrent protective device in the circuit concerned is available. A selection of such characteristics for fuses and circuit-breakers is given in Appendix 3 of BS 7671. For other types of device the advice of the manufacturer has to be sought. The time/current characteristics given in Appendix 3 indicate the values of I_a against various disconnection times for the devices tabulated and give co-ordinates for fixed times. For circuit-breakers, the curves indicate the maximum operating current, I_a, of a particular type of circuit-breaker and RCBO (see Table 3.4, section 3.5 of this Guidance Note for CB type details). However, for HRC fuses, the curves given in Appendix 3 are drawn on the median of the 'windows' provided for each rating in the fuse standard. Appx 3

Assuming that the size and type of cables to be used have already been determined from consideration of other aspects, such as mechanical protection and installation requirements, together with the magnitude of the design current of the circuit and the limitation of voltage drop under normal load conditions, the first stage is to calculate the earth fault loop impedance, Z_S. If the cable does not incorporate a suitable protective conductor, that conductor has to be chosen separately. Regulation 543.2.2 details what may be used as a protective conductor. 543.2.2

For cables having conductors of cross-sectional area not exceeding 35 mm², their inductive reactance can be ignored so that where these cables are used in radial circuits, the earth fault loop impedance Z_s is given by:

$$Z_s = Z_e + (R_1 + R_2)\ \Omega \qquad (1)$$

where:

Z_e is that part of the earth fault loop impedance external to the circuit concerned,
R_1 is the resistance of the circuit line conductor from the origin of the circuit where Z_e is taken, to the most distant socket-outlet or other point of utilization at normal conductor operating temperature, and
R_2 is the resistance of the circuit protective conductor from the origin of the circuit where Z_e is taken, to the most distant socket-outlet or other point of utilization at normal conductor operating temperature.

Similarly, where such cables are used in a ring final circuit without spurs, the earth fault loop impedance Z_s is given by:

$$Z_s = Z_e + (0.25\ R_1) + (0.25\ R_2)\ \Omega \qquad (2)$$

where:

Z_e is as described in **1** above,
R_1 is now the total resistance of the line conductor, in ring formation, between its ends prior to them being connected together to complete the ring, and
R_2 is similarly the total resistance of the protective conductor in ring formation.

(Again, R_1 and R_2 are taken at normal conductor operating temperature.)

Note: Strictly, the above equations are vectorial but arithmetic addition to determine the earth fault loop impedance may be used, as it gives a pessimistically high value for that impedance.

Having determined Z_s, the operating current I, i.e. the earth fault current I_f, is given by:

$$I_f = \frac{U_0}{Z_s} \text{ amperes} \qquad (3)$$

where U_0 is the nominal a.c. rms line voltage to Earth.

From the relevant overcurrent protective device time/current characteristic the time (t) for disconnection corresponding to this earth fault current is obtained.

Substituting for I_f, t and the appropriate k value in the equation

$$S = \frac{\sqrt{I_f^2\, t}}{k} \qquad (4)$$

where S is the circuit protective conductor minimum cross-sectional area in mm²

(**Note:** This equation is an adiabatic equation and is applicable for disconnection times not exceeding 5 s.)

543.1.3 from Regulation 543.1.3, this calculation then gives the minimum cross-sectional area of the protective conductor that will provide adequate thermal capacity to withstand the heat produced by the passage of fault current without damage. This has to be equal to or less than the circuit protective conductor size originally chosen (which has resistance R_2). This calculation is usually known as the adiabatic calculation. Where the line and protective conductors comply with the minimum size requirements of Table 54.7 of BS 7671 this adiabatic calculation is not necessary, as the conductors comply with the requirements of Regulation 543.1.4. 543.1.4

In order to assist the designer, when the cables selected for use are to Table 5 of BS 6004, or Table 7 of BS 7211, or are other thermoplastic (PVC) insulated and sheathed cables to the relevant British Standards, Tables D.1 to D.3 give the maximum earth fault loop impedances (Z_s) for circuits having line and protective conductors of copper from 1 mm^2 to 16 mm^2 cross-sectional area and where the overcurrent protective device is a fuse to BS 88-2.2, BS 88-6, BS 1361 or BS 3036. The tables also apply if the protective conductor is bare copper and in contact with cable insulated with thermoplastic (PVC).

For each type of fuse, two tables are given:

i where the circuit concerned is a final circuit not exceeding 32 A and the disconnection time for compliance with Regulation 411.3.2.2 is 0.4 s, and

ii where the circuit concerned is a distribution circuit or is a final circuit exceeding 32 A, supplied from a TN system, and the disconnection time for compliance with Regulation 411.3.2.3 is 5 s.

Table D.4 gives the maximum earth fault loop impedances for circuits where the overcurrent protective device is a circuit-breaker to BS 3871 Part 1 (now withdrawn) or BS EN 60898. The values given apply to both 0.4 s and 5 s disconnection times since, in practice, the overcurrents corresponding to the 0.4 s and 5 s disconnection times both cause the circuit-breaker to operate within 0.1 s.

It should be noted that circuit-breakers will operate within 0.1 s whenever the fault current is equal to or greater than the upper limit of their 'instantaneous' tripping band. That is:

- For Type B circuit-breakers – fault current $\geq 5\ I_n$
- For Type C circuit-breakers – fault current $\geq 10\ I_n$
- For Type D circuit-breakers – fault current $\geq 20\ I_n$.

Where the adiabatic calculation of S for compliance with Regulation 543.1.3 is carried out, the value of t in equation (4) is 0.1.

For circuits protected by circuit-breakers, compliance with Table D.4 provides compliance with Regulation 543.1.1 where the protective conductors range from 1 mm^2 to 16 mm^2 cross-sectional area and the rated current of the circuit-breakers ranges from 5 A to 125 A.

In each table the earth fault loop impedances given correspond to the appropriate disconnection time from a comparison of the time/current characteristic of the device concerned and the equation given in Regulation 543.1.3.

The tabulated values apply only when the nominal voltage to Earth (U_0) is 230 V.

For guidance on conductor resistances, see Appendix E.

The tabulated loop impedances are design figures assuming final conductor temperatures appropriate to earth fault conditions. For testing purposes the loop impedances must be reduced. For example, if testing at an ambient of 10 °C using cables to Table 5 of BS 6004, the maximum permitted test loop impedance is given by:

$$Z_{test} \leq (Z_{FX} - Z_e)\frac{0.96}{1.20} + Z_e \qquad (5)$$

where:

Z_{FX} is the loop impedance given by Tables D.1, D.2, D.3 or D.4
Z_e is the supply earth loop impedance
1.20 is the multiplier from Table E.2
0.96 is the ambient correction factor from Table E.3

▼ **Table D.1** Maximum earth fault loop impedance (in ohms) when overcurrent protective device is a semi-enclosed fuse to BS 3036

i 0.4 s disconnection (final circuits in TN systems, 32 A or less)

Protective conductor (mm²)	Fuse rating (A)			
	5	15	20	30
1.0	9.58	2.55	1.77	NP
1.5	9.58	2.55	1.77	1.1
2.5	9.58	2.55	1.77	1.1
4.0	9.58	2.55	1.77	1.1
6.0	9.58	2.55	1.77	1.1
10.0	9.58	2.55	1.77	1.1
16.0	9.58	2.55	1.77	1.1

ii 5 s disconnection (final circuits exceeding 32 A and distribution circuits in TN systems)

Protective conductor (mm²)	Fuse rating (A)			
	15	20	30	45
1.0	5.35	3.35*	NP	NP
1.5	5.35	3.83	2.49*	NP
2.5	5.35	3.83	2.64	1.43*
4.0	5.35	3.83	2.64	1.59
6.0	5.35	3.83	2.64	1.59
10.0	5.35	3.83	2.64	1.59
16.0	5.35	3.83	2.64	1.59

Notes:

▶ A value for k of 115 from Table 54.3 of BS 7671 is used for this table. This is also suitable for cables to Table 5 of BS 6004.

* Limited by thermal considerations of the protective conductor cross-sectional area, see Regulation 543.1.3.

NP Protective conductor, fuse combination not permitted.

▼ Table D.2 Maximum earth fault loop impedance (in ohms) when overcurrent protective device is a fuse to BS 88-2.2 or BS 88-6

i 0.4 s disconnection (final circuits in TN systems, 32 A or less)

Protective conductor (mm²)	Fuse rating (A)				
	6	10	16	20	32
1.0	8.52	5.11	2.70	1.77	0.82*
1.5	8.52	5.11	2.70	1.77	1.04
2.5	8.52	5.11	2.70	1.77	1.04
4.0	8.52	5.11	2.70	1.77	1.04
6.0	8.52	5.11	2.70	1.77	1.04
10.0	8.52	5.11	2.70	1.77	1.04
16.0	8.52	5.11	2.70	1.77	1.04

ii 5 s disconnection (final circuits exceeding 32 A and distribution circuits in TN systems)

Protective conductor (mm²)	Fuse rating (A)						
	6	10	16	20	32	40	50
1.0	13.5	7.42	3.82*	2.09*	0.80	NP	NP
1.5	13.5	7.42	4.18	2.91	1.17	NP	NP
2.5	13.5	7.42	4.18	2.91	1.84	1.03*	0.68*
4.0	13.5	7.42	4.18	2.91	1.84	1.35	0.96*
6.0	13.5	7.42	4.18	2.91	1.84	1.35	1.04
10.0	13.5	7.42	4.18	2.91	1.84	1.35	1.04
16.0	13.5	7.42	4.18	2.91	1.84	1.35	1.04

Notes:

▶ A value for k of 115 from Table 54.3 of BS 7671 is used for this table. This is also suitable for cables to Table 5 of BS 6004.

* Limited by thermal considerations of the protective conductor cross-sectional area, see Regulation 543.1.3.

NP Protective conductor, fuse combination not permitted.

Table D.3
Maximum earth fault loop impedance (in ohms) when overcurrent protective device is a fuse to BS 1361

i 0.4 s disconnection (final circuits in TN systems, 32 A or less)

Protective conductor (mm^2)	Fuse rating (A)			
	5	15	20	30
1.0	10.45	3.28	1.70	0.96*
1.5	10.45	3.28	1.70	1.15
2.5	10.45	3.28	1.70	1.15
4.0	10.45	3.28	1.70	1.15
6.0	10.45	3.28	1.70	1.15
10.0	10.45	3.28	1.70	1.15
16.0	10.45	3.28	1.70	1.15

ii 5 s disconnection (final circuits exceeding 32 A and distribution circuits in TN systems)

Protective conductor (mm^2)	Fuse rating (A)				
	5	15	20	30	45
1.0	16.4	5.00	2.09*	0.96*	NP
1.5	16.4	5.00	2.68*	1.43*	0.41*
2.5	16.4	5.00	2.80	1.84	0.63*
4.0	16.4	5.00	2.80	1.84	0.88*
6.0	16.4	5.00	2.80	1.84	0.96
10.0	16.4	5.00	2.80	1.84	0.96
16.0	16.4	5.00	2.80	1.84	0.96

Notes:

- A value for k of 115 from Table 54.3 of BS 7671 is used for this table. This is also suitable for cables to Table 5 of BS 6004.

* Limited by thermal considerations of the protective conductor cross-sectional area, see Regulation 543.1.3.

NP Protective conductor, fuse combination not permitted.

Table D.4i Maximum earth fault loop impedance (in ohms) when overcurrent protective devices are circuit-breakers to BS EN 60898 or RCBOs to BS EN 61009-1

Table 41.3
411.4.7

Both 0.4 and 5 s disconnection times (notes 1 and 2)

Circuit-breaker type	Circuit-breaker rating (A)											
	6	10	16	20	25	32	40	50	63	80	100	125
B	7.67	4.60	2.87	2.30	1.84	1.44	1.15	0.92	0.73	0.57	0.46	0.37
C	3.83	2.30	1.44	1.15	0.92	0.72	0.57	0.46	0.36	0.29	0.23	0.18
D	1.92	1.15	0.72	0.57	0.46	0.36	0.29	0.23	0.18	0.14	0.11	0.09

Table D.4ii Maximum earth fault loop impedance (in ohms) when overcurrent protective devices are magnetic circuit-breakers to BS 3871

Both 0.4 and 5 s disconnection times (note 1)

Circuit-breaker type	Circuit-breaker rating (A)												
	5	6	10	15	16	20	25	30	32	40	45	50	63
1	11.50	9.58	5.75	3.83	3.59	2.87	2.30	1.92	1.80	1.44	1.27	1.15	0.91
2	6.57	5.47	3.29	2.19	2.05	1.64	1.31	1.09	1.02	0.82	0.73	0.66	0.52
3	4.60	3.83	2.30	1.53	1.44	1.15	0.92	0.77	0.72	0.57	0.51	0.46	0.36

Notes:

1 A value for k of 115 from Table 54.3 of BS 7671 is used. This is suitable for thermoplastic (PVC) insulated and sheathed cables to Table 5 and Table 6 of BS 6004 and for lshf (low smoke halogen-free) insulated and sheathed cables to Table 7 of BS 7211. The k value is based on both the thermoplastic (PVC) and lshf cables operating at a maximum operating temperature of 70 °C.

2 In TN systems it is preferable for reliable operation for fault protection to be provided by overcurrent devices, including RCBOs operating as overcurrent devices; that is, with loop impedance complying with the table above. RCBOs are then providing fault protection as circuit-breakers and additional protection as RCDs. If, however, the residual current element of an RCBO has been selected to provide fault protection, the tabulated values of maximum earth fault loop impedance do not apply. 411.4 411.3.3

D.1 Selection of a circuit protective conductor

As previously stated, Regulation 543.1.1 allows the selection of a circuit protective conductor (cpc) by calculation in accordance with Regulation 543.1.3 or by reference to Table 54.7 to select the minimum size, in accordance with Regulation 543.1.4.

Table 54.7 provides a method to establish the minimum size of cpc based on the area of the line conductor of the circuit. In many cases the material and size of the intended cpc is established by the selection of the wiring system to be used – i.e. steel-wire armoured cables, steel conduit, mineral insulated cables, etc. If the cross-sectional area of the armour, conduit, etc., is found to comply with the requirements of Table 54.7 then calculation of the thermal capacity is not usually necessary.

D.2 Steel conduit and trunking

For some types of cpc, e.g. steel conduit, a check to prove compliance with Table 54.7 is quite simple and minimum cross-sectional areas of heavy gauge steel conduit and steel trunkings are listed in Table D.5. Taking the value of k_1 as 115 for thermoplastic (PVC) from Table 43.1 and k_2 as 47 from Table 54.5 it can be shown that heavy gauge steel conduit is a suitable protective conductor for all sizes of conductor that can be drawn into it in accordance with Appendix A.

A well installed and maintained steel conduit or cable trunking system can provide an adequate cpc for a final circuit (Regulation 543.2.2). Unfortunately, conduit and trunking impedance data for the calculation of earth fault loop impedances of circuits utilizing conduit or trunking as a protective conductor is not readily available and so the practice of installing a separate cpc has arisen. This is usually uneconomical and unnecessary.

However, when a conduit or trunking is to be utilized as a protective conductor, all joints must be tight and secure and the system protected from corrosion. The cross-sectional area of the protective conductor is to be taken as the minimum of any variations of size along the route. Consideration must also be made of the possible further reduction in cross-sectional area at conduit running couplers and cable trunking couplers, with only metal-to-metal contact at bolt holes even when copper links are fitted. Flanged couplers may be necessary to provide adequate continuity at conduit connections onto trunking systems. It must be remembered that the Electricity at Work Regulations 1989 require that an adequate connection to earth be provided where this is needed to prevent danger (Regulation 8).

Table D.5
Cross-sectional areas of steel conduit and trunking

Heavy Gauge Steel Conduit to BS 4568-1 (BS EN 61386-21)

Nominal diameter (mm)	Minimum steel cross-sectional area (mm^2)
16	64.4
20	82.6
25	105.4
32	137.3

Steel Surface Trunking to BS EN 50085-1 (sample sizes)

Nominal size (mm x mm)	Minimum steel cross-sectional area without lid (mm^2)
50 x 50	135
75 x 75	243
100 x 50	216
100 x 100	324
150 x 100	378

Steel Underfloor Trunking to BS 4678-2 (sample sizes)

Nominal size (mm x mm)	Minimum steel cross-sectional area without lid (mm^2)
75 x 25	118
100 x 50	142
100 x 100	213
150 x 100	284

The requirements of Regulation 543.3, as they apply, must be complied with in respect of the preservation of electrical continuity, for a protective conductor common to several circuits. The cross-sectional area must either be calculated to provide adequate capacity for the most onerous conditions of any of the several circuits, or sized from Table 54.7 to correspond with the largest line conductor of the several circuits. Therefore, a metal conduit or trunking could serve as a cpc to all the group of circuits it encloses.

Formulae for the calculation of the resistance and inductive reactance values of steel conduit and steel ducting and trunking are published in Guidance Note 6.

D.3 Steel-wire armoured cables

If a BS EN 60269-2 or BS 88 fuse or a circuit-breaker is selected for overload protection of thermoplastic (PVC)/SWA/thermoplastic (PVC) cable to BS 6346 or XLPE/SWA/thermoplastic (PVC) cable to BS 5467 then for an earth fault disconnection time of up to 5 s the steel armouring will have a sufficient cross-sectional area to comply with Regulation 543.1.3 with two exceptions. These exceptions are two-core 240 mm^2 and 300 mm^2 XLPE/SWA/thermoplastic (PVC) to BS 5467.

From I, t and k, the required minimum cross-sectional area of wire armour required, S, can be calculated and compared with the area provided by the cable. The cross-sectional areas of steel-wire armour on thermosetting insulated cables to BS 5467 and thermoplastic insulated cables to BS 6346 are given in Tables 12E and 12F of the *Commentary on IEE Wiring Regulations*.

It will be found that, if the protective device is one of the standard types mentioned in the Regulations and if its rating is not greater than the current rating of the cable, the cross-sectional area of the armour is greater than required to comply with the Regulations. This assumes that the earth fault loop impedance of the circuit is such as to ensure that the device will operate in not more than 5 s on the occurrence of a fault of negligible impedance.

▼ **Table D.6**
Copper conductor, thermoplastic (PVC) insulated steel-wire armoured cables to BS 6346 and fuses to BS 88

Fuse rating	Required area of SWA (5 s disconnection)	Smallest cable providing required area of SWA			
		2-core	3-core	4-core (equal)	4-core (red. neut.)
A	mm^2	mm^2	mm^2	mm^2	mm^2
*63	14.3	1.5 (20 A)	1.5 (18 A)	1.5 (18 A)	–
80	21	4 (37 A)	4 (31 A)	2.5 (24 A)	–
100	28	10 (69 A)	6 (42 A)	4 (31 A)	–
125	35	10 (69 A)	6 (42 A)	4 (31 A)	–
160	46	16 (90 A)	16 (77 A)	10 (58 A)	–
200	61	25 (121 A)	25 (102 A)	16 (77 A)	–
250	77	70 (220 A)	70 (220 A)	50 (155 A)	35 (125 A)
315	107	95 (270 A)	70 (190 A)	50 (155 A)	70 (190 A)
400	133	120 (310 A)	95 (230 A)	70 (190 A)	70 (190 A)
500	178	185 (410 A)	150 (310 A)	120 (310 A)	120 (270 A)
630	224	240 (485 A)	185 (350 A)	150 (310 A)	150 (310 A)
800	350	–	–	400 (550 A)	400 (550 A)

* For BS EN 60269-2 or BS 88 fuses of 6–50 A ratings the required armour area is, of course, less than that required for the 63 A fuse, and the cables with 1.5 mm^2 conductors (the smallest in BS 6346) provide areas exceeding the required minima.

▼ **Table D.7**
Copper conductor, thermoplastic (PVC) insulated steel-wire armoured cables to BS 6346 and circuit-breakers to BS 3871 and BS EN 60898 (XLPE cables to BS 5467 also comply with the same requirements)

Circuit-breaker designation	Required area of SWA (5 s disconnection)	Smallest cable providing required area of SWA		
		2-core	3-core	4-core
	mm^2	mm^2	mm^2	mm^2
*30/32 A Type 1	6.1	1.5 (20 A)	1.5 (18 A)	1.5 (18 A)
30/32 A Type 2 or Type B	10.7	1.5 (20 A)	1.5 (18 A)	1.5 (18 A)
30/32 A Type 3 or Type C	13.7	1.5 (20 A)	1.5 (18 A)	1.5 (18 A)
50 A Type 1	10.2	1.5 (20 A)	1.5 (18 A)	1.5 (18 A)
50 A Type 2 or Type B	17.8	1.5 (20 A)	1.5 (18 A)	1.5 (18 A)
50 A Type 3 or Type C	21.8	4 (37 A)	4 (31 A)	4 (31 A)

* For 5–20 A circuit-breakers of all types to BS 3871 and 6–25 A for BS EN 60898, the 1.5 mm^2 cables provide armour areas exceeding the required minima.

Fuse rating	Required area of SWA (5 s disconnection)	Smallest cable providing required area of SWA			
		2-core	3-core	4-core (equal)	4-core (red. neut.)
A	mm^2	mm^2	mm^2	mm^2	mm^2
*125	28.2	16 (114 A)	16 (100 A)	16 (100 A)	–
160	37	16 (114 A)	16 (100 A)	16 (100 A)	–
200	50	35 (190 A)	25 (133 A)	25 (133 A)	25 (133 A)
250	62	50 (228 A)	25 (133 A)	25 (133 A)	25 (133 A)
315	87	95 (356 A)	70 (247 A)	50 (195 A)	50 (195 A)
400	108	95 (356 A)	95 (304 A)	70 (247 A)	70 (247 A)
500	145	185 (540 A)	150 (408 A)	95 (304 A)	95 (304 A)
630	182	185 (540 A)	150 (408 A)	120 (350 A)	150 (408 A)
800	286	–	–	240 (550 A)	300 (627 A)

▼ **Table D.8**
Copper conductor, 90 °C thermosetting insulated steel-wire armoured cables to BS 5467 and fuses to BS 88

* For fuses to BS EN 60269-2 or BS 88 of lower rating than 125 A, cables with 16 mm^2 conductors (the smallest size in BS 5467) provide armour areas exceeding the required minima.

Tables are also given below identifying which of the standard sizes of steel-wire armoured cables can be utilized to comply with Table 54.7.

Conductor csa (mm^2)	Minimum csa of steel cpc required to comply with Table 54.7 (mm^2)	Actual armour csa from BS 6346 (mm^2)		
		2-core	3-core	4-core
1.5	3.4	15	16	17
2.5	5.7	17	19	20
4	9.0	21	23	35
6	13.6	24	36	40
10	22.6	41	44	49
16	36.1	46	50	72
25	36.1	60	66	76
35	36.1	66	74	84
50	56.4	74	84	122
70	79.0	84	119	138
95	107.2	122	138	160
120	135.3	~~(131)~~	150	220
150	169.2	~~(144)~~	211	240
185	208.6	~~(201)~~	230	265
240	270.6	~~(225)~~	~~(260)~~	299
300	338.3	~~(250)~~	~~(289)~~	~~(333)~~
400	403.9	~~(279)~~	~~(319)~~	467

▼ **Table D.9**
Thermoplastic (PVC)/SWA/thermoplastic cables to BS 6346. Insulation operating at 70 °C.

For Table 54.7, $k_1 = 115$ (103 for 400 mm^2) from Table 43.1 and $k_2 = 51$ from Table 54.4

~~( )~~ Indicates that the cable does not comply with Table 54.7. Therefore, the cpc size must be confirmed by calculation as indicated by Regulation 543.1.3, or a supplementary cpc of the full conductor size must be installed.

▼ **Table D.10a**
90 °C thermosetting insulated steel-wire armoured cables to BS 5467 and BS 6724. Insulation operating at 90 °C.

For Table 54.7, $k_1 = 143$ from Table 43.1 and $k_2 = 46$ from Table 54.4

Conductor csa (mm²)	Minimum csa of steel cpc required to comply with Table 54.7 (mm²)	Actual armour csa from BS 5467 and BS 6724 (mm²)		
		2-core	3-core	4-core
1.5	4.7	15	16	17
2.5	7.8	17	19	20
4	12.5	19	21	23
6	18.7	22	23	36
10	31.1	~~(26)~~	39	43
16	49.8	~~(41)~~	~~(44)~~	~~(49)~~
25	49.8	~~(42)~~	62	70
35	49.8	62	70	80
50	77.8	~~(68)~~	78	90
70	108.8	~~(80)~~	~~(90)~~	131
95	147.7	~~(113)~~	~~(128)~~	~~(147)~~
120	186.6	~~(125)~~	~~(141)~~	206
150	233.2	~~(138)~~	~~(201)~~	~~(230)~~
185	287.6	~~(191)~~	~~(220)~~	~~(255)~~
240	373.1	~~(215)~~	~~(250)~~	~~(289)~~
300	466.3	~~(235)~~	~~(269)~~	~~(319)~~
400	521.7	~~(265)~~	~~(304)~~	~~(452)~~*

* Not applicable to BS 6724.

~~( )~~ Indicates that the cable does not comply with Table 54.7. Therefore, the cpc size must be confirmed by calculation as indicated by Regulation 543.1.3, or a supplementary cpc of the full conductor size must be installed.

There are times when an XLPE insulated cable, normally rated at 90 °C, may be run alongside thermoplastic (PVC) insulated cables operating at 70 °C. Under such circumstances the 90 °C thermosetting insulated cable must be limited to the lower operating temperature of 70 °C and Table D.10b should be used to determine if the armour complies with Table 54.7.

Table D.10b
90 °C thermosetting insulated steel-wire armoured cables to BS 5467 and BS 6724. Insulation operating at 70 °C.

For Table 54.7, $k_1 = 115$ from Table 43.1 and $k_2 = 51$ from Table 54.4

Conductor csa (mm²)	Minimum csa of steel cpc required to comply with Table 54.7 (mm²)	Actual armour csa from BS 5467 and BS 6724 (mm²)		
		2-core	3-core	4-core
1.5	3.4	16	17	18
2.5	5.7	17	19	20
4	9.0	19	21	23
6	13.6	22	23	36
10	22.6	26	39	43
16	36.1	41	44	49
25	36.1	42	62	70
35	36.1	62	70	80
50	56.4	68	78	90
70	79.0	80	90	131
95	107.2	113	128	147
120	135.3	~~(125)~~	141	206
150	169.2	~~(138)~~	201	230
185	208.6	~~(191)~~	220	255
240	270.6	~~(215)~~	~~(250)~~	289
300	388.3	~~(235)~~	~~(269)~~	~~(319)~~
400	451.0	~~(265)~~	~~(304)~~	452*

* Not applicable to BS 6724.

~~( )~~ Indicates that the cable does not comply with Table 54.7. Therefore, the cpc size must be confirmed by calculation as indicated by Regulation 543.1.3 or a supplementary cpc of the full conductor size must be installed.

D.4 Mineral insulated cable with copper sheath

Cross-sectional areas of the copper sheath of light duty and heavy duty mineral insulated cables to BS EN 60702-1 are given in Table D.11, in which compliance with Table 54.7 is indicated.

▼ **Table D.11**
Cross-sectional areas of the copper sheath of light and heavy duty mineral cables to BS EN 60702-1

Cable size reference	Effective sheath area (mm^2)
500 V Grade (light duty)	
2L1	5.4
2L1.5	6.3
2L2.5	8.2
2L4	10.7
3L1	6.7
3L1.5	7.8
3L2.5	9.5
4L1	7.7
4L1.5	9.1
4L2.5	11.3
7L1	10.2
7L1.5	11.8
7L2.5	15.4
750 V Grade (heavy duty)	
1H6.0	8.0
1H10	~~(9.0)~~
1H16	~~(12.0)~~
1H25	~~(15.0)~~
1H35	~~(18.0)~~
1H50	~~(22.0)~~
1H70	~~(27.0)~~
1H95	~~(32.0)~~
1H120	~~(37.0)~~
1H150	~~(44.0)~~
1H185	~~(54.0)~~
1H240	~~(70.0)~~
2H1.5	11
2H2.5	13
2H4	16
2H6	18
2H10	24
2H16	30
2H25	38
3H1.5	12
3H2.5	14
3H4	17
3H6	20
3H10	27
3H16	34

continues

Cable size reference	Effective sheath area (mm²)
3H25	42
4H1.5	14
4H2.5	16
4H4	20
4H6	24
4H10	30
4H16	39
4H25	49
7H1.5	18
7H2.5	22
12H2.5	34
19H2.5	37

▼ **Table D.11** *continued*

(⟶) Indicates that the cable does not comply with Table 54.7. Therefore, the cpc size must be confirmed by calculation as indicated by Regulation 543.1.3, or a supplementary cpc must be installed.

It can be seen that all multicore sizes comply with Table 54.7 but, with the exception of 1H6, the heavy duty single-core micc cables do not. Although these cables singularly do not comply, usually single-core cables would be run in pairs for a single-phase circuit, or three or four for a three-phase circuit. The sheath sections, if bonded together, would then sum together and comply for some conductors. (See Regulation 523.10 also.) 523.10

D.5 Thermoplastic (PVC) insulated and sheathed cables to BS 6004

These common wiring cables have a cpc of a reduced size from that of the line conductor for all sizes excepting 1.0 mm². Consequently, only the 1.0 mm² cables (6241Y, 6242Y and 6243Y) will comply with Table 54.7. The standard conductor sizes of these cables are identified in Table E.1 of Appendix E.

Consequent upon this, calculations are required for the cpc in circuits in which this type of cable is used, usually in domestic installations. The *On-Site Guide* contains Table 7.1 giving calculated maximum circuit lengths for various arrangements of standard circuits and types of protective device.

Similar cable size constructions apply to thermosetting cables to Table 7 of BS 7211: 1998. These cables, however, can have a conductor operating temperature of up to 90 °C. Table 52.1 of BS 7671 requires that equipment and accessories connected to such conductors operating at temperatures exceeding 70 °C must be suitable for the resulting higher temperature at the connection. Table 52.1

Domestic accessories and equipment are not usually suitable for operation at such higher temperatures and consequently cables to BS 7211 cannot be used at their full rating and should be considered as thermoplastic (PVC) insulated and sheathed types.

Resistance and impedance of copper and aluminium conductors under fault conditions

E

To check compliance with Regulation 434.5.2 and/or Regulation 543.1.3, i.e. to evaluate the equation $S^2 = I^2 t/k^2$, it is necessary to establish the impedances of the circuit conductors to determine the fault current I and hence the protective device disconnection time t.

Similarly, in order to design circuits for compliance with BS 7671, the limiting values of earth fault loop impedance given in Tables 41.2 to 41.4 of BS 7671, it is necessary to establish the relevant impedances of the circuit conductors concerned. (See Guidance Note 5 for more information on this subject.)

Where the circuit overcurrent protective device characteristics comply with those given in Appendix 3 of BS 7671, the increase in circuit conductor temperature at fault has been deemed to be taken into account and only the normal conductor operating temperature needs to be considered.

The equation given in Regulations 434.5.2 and 543.1.3 has been based on the assumption of a constant value of fault current but in practice that current changes during the period of the fault because, due to the rise in temperature, the conductor resistance increases.

The rigorous method for taking into account the changing character of the fault current is too complicated for practical use and over the range of temperatures encountered in BS 7671 a sufficiently accurate method is to calculate conductor impedances based on the full load current temperature of the conductor (i.e. 70 °C for thermoplastic insulated conductors and 90 °C for thermosetting insulated conductors) for devices given in Appendix 3 of BS 7671.

Appendix 3 does not include all types of device, for example circuit-breakers (MCCBs) to BS EN 60947-2:2006. Earth fault loop impedances are not tabulated in BS 7671 for MCCBs nor are characteristics shown in Appendix 3. It may be that these devices do comply with the current regulation requirements but there is a wide variety of devices with adjustable ranges and characteristics and if there is any doubt, the temperature rise of the average of the initial and final fault temperatures should be utilized. The device manufacturer's advice may be required.

Most manufacturers will, however, provide details of the maximum allowable impedance values for both 0.4 s and 5 s disconnection.

The following conductor resistance Table E.1 is limited to conductor cross-sectional areas up to and including 35 mm^2, i.e. to conductors having negligible inductive reactance. For larger cables the reactance is not negligible. This reactance is independent of temperature and depends on conductor size and cable make-up and it may be necessary to obtain information from the manufacturer as regards the resistive and reactive components of the impedance of the cables it is intended to use.

Alternatively, values can be calculated for some cables from the voltage drop data given in Appendix 4 of BS 7671. Voltage drop data is given in terms of mV/A/m (which is mΩ/m) at full load conductor temperature. Single-phase figures give the resistance/ reactance/impedance voltage drop for both line and neutral conductors, which must be divided by two. Three-phase resistance/reactance/impedance voltage drop values are given for a balanced three-phase conductor arrangement. The figures must be divided by $\sqrt{3}$ (1.732) to give a figure for one line conductor only. It should be noted that the reactive component, and consequently the impedance, also varies significantly depending on the cable installation conditions and method. The correct details must be chosen.

Table E.1 gives values of (R_1 + R_2) per metre for various combinations of sizes of conductors up to and including 35 mm^2 cross-sectional area (see Appendix D for descriptions of R_1 and R_2). It also gives values of resistance per metre for each size of conductor.

Table E.2 gives the multipliers to be applied to the values given in Table E.1 for the purpose of calculating the resistance, at maximum permissible operating temperature, of the line conductors and/or circuit protective conductors, in order to determine compliance with, as applicable:

1 earth fault loop impedances of Tables 41.2, 41.3 or 41.4
2 the equation in Regulation 434.5.2
3 the equation in Regulation 543.1.3
4 earth fault loop impedances of Appendix D in this Guidance Note.

Where it is known that the actual operating temperature under normal load is less than the maximum permissible value for the type of cable insulation concerned (as given in the tables of current-carrying capacity), the multipliers given in Table E.2 may be reduced. (See Appendix 4 of BS 7671 for further details.)

Table E.1
Values of resistance/metre for copper and aluminium conductors and of $(R_1 + R_2)$/metre in milliohms/metre

Cross-sectional area (mm²)		Resistance/metre or $(R_1 + R_2)$/metre (mΩ/m) at 20 °C		Resistance/metre or $(R_1 + R_2)$/metre (mΩ/m) at 70 °C	
Line conductor	Protective conductor	Plain copper	Aluminium	Plain copper	Aluminium
1	–	18.10		21.72	
1*	1	36.20		43.44	
1.5	–	12.10		14.52	
1.5*	1	30.20		36.24	
1.5	1.5	24.20		29.04	
2.5	–	7.41		8.89	
2.5	1	25.51		30.61	
2.5*	1.5	19.51		23.41	
2.5	2.5	14.82		17.78	
4	–	4.61		5.53	
4*	1.5	16.71		20.05	
4	2.5	12.02		14.42	
4	4	9.22		11.06	
6	–	3.08		3.70	
6*	2.5	10.49		12.59	
6	4	7.69		9.23	
6	6	6.16		7.39	
10	–	1.83		2.20	
10*	4	6.44		7.73	
10	6	4.91		5.89	
10	10	3.66		4.39	
16	–	1.15	1.91	1.38	2.29
16*	6	4.23	–	5.08	–
16	10	2.98	–	3.58	–
16	16	2.30	3.82	2.76	4.58
25	–	0.727	1.20	0.87	1.44
25	10	2.557	–	3.07	–
25	16	1.877	–	2.25	–
25	25	1.454	2.40	1.74	2.88
35	–	0.524	0.868	0.63	1.04
35	16	1.674	2.778	2.01	3.33
35	25	1.251	2.068	1.50	2.48
35	35	1.048	1.736	1.26	2.08

* Identifies copper line/protective conductor arrangement that complies with Table 5 or Table 6 of BS 6004:2000 for thermoplastic insulated and sheathed single, twin or three-core and cpc cables (i.e. 6241Y, 6242Y or 6243Y cables) and similar cable constructions for thermosetting cables to BS 7211:1998.

▼ **Table E.2**
Operating temperature multipliers to Table E.1

Insulation material	Multiplier	
	54.2	54.3
70 °C thermoplastic	1.04	1.20
90 °C thermoplastic	1.04	1.28
90 °C thermosetting	1.04	1.28

The multipliers given in Table E.2 are based on the simplified formula given in BS 6360 for both copper and aluminium conductors, namely, that the resistance temperature coefficient, $\propto$, is 0.004 per °C at 20 °C. This is $R = R_{20}\ [1 + \propto \Theta]$ where Θ is the temperature in degrees Celsius between the initial 20 °C ambient and the final full load conductor temperature and R_{20} is the conductor resistance at 20 °C.

Table 54.2 **54.2** applies where the protective conductor is not incorporated or bunched with cables, or for bare protective conductors in contact with cable covering (assumed initial temperature 30 °C).

Table 54.3 **54.3** applies where the protective conductor is a core in a cable or is bunched with cables (assumed initial temperature 70 °C or greater).

E.1 Verification

For verification purposes of earth fault loop impedances at installation completion the person carrying out testing will need the maximum values of the line and protective conductor resistances of a circuit at the ambient temperature expected during tests. This may be different from the reference temperature of 20 °C used for Table E.1.

Table E.3 gives correction factors that may be applied to the 20 °C Table E.1 values to take account of the ambient temperature (for test purposes only). See also Guidance Note 3: *Inspection & Testing.*

▼ **Table E.3**
Ambient temperature multipliers to Table E.1

Expected ambient temperature (°C)	Correction factor
5	0.94
10	0.96
15	0.98
25	1.02

Tables have been published in the *On-Site Guide* to provide maximum circuit earth fault loop impedances at 20 °C ready for test comparison (see Appendix 2 of the *On-Site Guide*).

E.2 Mineral insulated cable with copper sheath

Values of conductor resistance (R_1) and sheath resistance (R_2) per metre for copper sheathed light duty and heavy duty mineral insulated cables to BS EN 60702 are given in Table E.4.

▼ **Table E.4**
Values of R_1 and R_2 (in Ω/km) for mineral insulated cable with copper sheath

Cable ref.	R_1 Conductor resistance 20 °C	R_2 Sheath resistance 20 °C	R_1 Conductor resistance	R_2 Sheath resistance	R_1 Conductor resistance	R_2 Sheath resistance
			Exposed to touch 70 °C sheath		Not exposed to touch 105 °C sheath	
500 V Grade (light duty)						
2L1	18.1	3.95	21.87	4.47	24.5	4.84
2L1.5	12.1	3.35	14.62	3.79	16.38	4.1
2L2.5	7.41	2.53	8.95	2.87	10.03	3.1
2L4	4.61	1.96	5.57	2.22	6.24	2.4
3L1	18.1	3.15	21.87	3.57	24.5	3.86
3L1.5	12.1	2.67	14.62	3.02	16.38	3.27
3L2.5	7.41	2.23	8.95	2.53	10.03	2.73
4L1	18.1	2.71	21.87	3.07	24.5	3.32
4L1.5	12.1	2.33	14.62	2.64	16.38	2.85
4L2.5	7.41	1.85	8.95	2.1	10.03	2.27
7L1	18.1	2.06	21.87	2.33	24.5	2.52
7L1.5	12.1	1.78	14.62	2.02	16.38	2.18
7L2.5	7.41	1.36	8.95	1.54	10.03	1.67
750 V Grade (heavy duty)						
1H10	1.83	2.23	2.21	2.53	2.48	2.73
1H16	1.16	1.81	1.4	2.05	1.57	2.22
1H25	0.727	1.4	0.878	1.59	0.984	1.72
1H35	0.524	1.17	0.633	1.33	0.709	1.43
1H50	0.387	0.959	0.468	1.09	0.524	1.18
1H70	0.268	0.767	0.324	0.869	0.363	0.94
1H95	0.193	0.646	0.233	0.732	0.261	0.792
1H120	0.153	0.556	0.185	0.63	0.207	0.681
1H150	0.124	0.479	0.15	0.542	0.168	0.587
1H185	0.101	0.412	0.122	0.467	0.137	0.505
1H240	0.0775	0.341	0.0936	0.386	0.105	0.418
2H1.5	12.1	1.9	14.62	2.15	16.38	2.33
2H2.5	7.41	1.63	8.95	1.85	10.03	2
2H4	4.61	1.35	5.57	1.53	6.24	1.65
2H6	3.08	1.13	3.72	1.28	4.17	1.38
2H10	1.83	0.887	2.21	1.005	2.48	1.09
2H16	1.16	0.695	1.4	0.787	1.57	0.852
2H25	0.727	0.546	0.878	0.618	0.984	0.669
3H1.5	12.1	1.75	14.62	1.98	16.38	2.14
3H2.5	7.41	1.47	8.95	1.66	10.03	1.8

continues

▼ **Table E.4**
continued

Cable ref.	R_1 Conductor resistance 20 °C	R_2 Sheath resistance 20 °C	R_1 Conductor resistance	R_2 Sheath resistance	R_1 Conductor resistance	R_2 Sheath resistance
			Exposed to touch 70 °C sheath		Not exposed to touch 105 °C sheath	
3H4	4.61	1.23	5.57	1.39	6.24	1.51
3H6	3.08	1.03	3.72	1.17	4.17	1.26
3H10	1.83	0.783	2.21	0.887	2.48	0.959
3H16	1.16	0.622	1.4	0.704	1.57	0.762
3H25	0.727	0.5	0.878	0.566	0.984	0.613
4H1.5	12.1	1.51	14.62	1.71	16.38	1.85
4H2.5	7.41	1.29	8.95	1.46	10.03	1.58
4H4	4.61	1.04	5.57	1.18	6.24	1.27
4H6	3.08	0.887	3.72	1	4.17	1.09
4H10	1.83	0.69	2.21	0.781	2.48	0.845
4H16	1.16	0.533	1.4	0.604	1.57	0.653
4H25	0.727	0.423	0.878	0.479	0.984	0.518
7H1.5	12.1	1.15	14.62	1.3	16.38	1.41
7H2.5	7.41	0.959	8.95	1.09	10.03	1.18
12H1.5	12.1	0.744	14.62	0.843	16.38	0.912
12H2.5	7.41	0.63	8.95	0.713	10.03	0.772
19H1.5	12.1	0.57	14.62	0.646	16.38	0.698

The calculation of $R_1 + R_2$ for mineral insulated cables is different from the method for other cables, in that the loaded conductor temperature is not usually given. Table 4G of Appendix 4 of BS 7671 gives normal full load sheath operating temperatures of 70 °C for thermoplastic sheathed types and 105 °C for bare cables not in contact with combustible materials, in a 30 °C ambient. Magnesium oxide is a relatively good thermal conductor and, being in a thin layer, it is found that conductor temperatures are usually only some 3 °C higher than sheath temperatures.

However, calculations are complicated as the sheath is of copper which is manufactured to a different material standard to that of the conductors and the $\propto$ factor of 0.004 at 20 °C for normal copper conductor temperature correction does not apply. An $\propto$ value of 0.00275 at 20 °C can be used for sheath resistance change calculations. Table E.4 gives calculated values of R_1 and R_2 at a standard 20 °C and at standard sheath operating temperatures, and these can be used directly for calculations at full load temperatures.

An estimate of the sheath temperature can be made for a partially loaded cable, in any higher ambient temperature, using the following formula.

$$\text{Approximate sheath temperature} = T_{amb} + \left[\left(\frac{I_b}{I_t}\right)^2 \times 40\right]$$

where:

T_{amb} is actual ambient temperature
I_b is circuit design current
I_t is tabulated current-carrying capacity of the cable.

E.3 Steel-wire armour, steel conduit and steel trunking

Formulae for the calculation of the resistance and inductive reactance values of the steel-wire armour of cables and of steel conduit, ducting and trunking are published in Guidance Note 6.

Generally, it is accepted that there is approximately a 10 °C difference between the conductor temperature and the outer sheath temperature for a steel-wire armoured cable at full load.

Selection and erection of wiring systems

F

F.1 General

To conform to the requirements of BS 7671, Regulation 511.1 requires that all electrical equipment, including wiring systems, must utilize cables complying with the relevant requirements of the applicable British Standard or Harmonized Standard. 511.1

Alternatively, if equipment complying with a foreign national standard based on an IEC Standard is to be used, the designer or other person responsible for specifying the installation must verify that any differences between that standard and the corresponding British Standard or Harmonized Standard will not result in a lesser degree of safety than that afforded by compliance with the British Standard.

Where equipment to be used is not covered by a British Standard or Harmonized Standard or is used outside the scope of its standard, the designer or other person responsible for specifying the installation must satisfy themselves and confirm that the equipment provides the same degree of safety as that afforded by compliance with the Regulations.

A 'wiring system' is defined in Part 2 of BS 7671 as 'An assembly made up of cable or busbars and parts which secure and, if necessary, enclose the cable or busbars'. This can be read to only mean factory-made systems but it is intended to cover all cable types. Part 2

Cables are also identified with a voltage grade, to identify the maximum system working voltage for which they are suitable. Conduit wiring cables (6491X), etc., are designated 450/750 V and are harmonized within CENELEC under HD 21 and HD 22. Wiring cables such as flat twin and earth (6242Y) are designated 300/500 V and are not harmonized but are constructed to a British Standard. Armoured cables are designated 600/1000 V and are not harmonized but are also constructed to a British Standard. There is no difference in utilizing types of any of these designations on the UK 230/400 V supply system.

BS 7540:2005 (series) is a guide to use for cables with a rated voltage not exceeding 450/750 V and gives installation application advice.

▼ **Table F.1** Comparison of harmonized cable types to BS 6004

Type	British Standard table no.	Code designation	Conductor class	Voltage grade (V)	Number of cores	csa range (mm^2)	Temperature (°C) Installation (minimum)	Temperature (°C) Storage (maximum)
Thermoplastic insulated non-sheathed	1	H07V-U H07V-R H07V-K	1 2 5	450/750	1	1.5–16 1.5–630 1.5–240	+5	+40
Thermoplastic insulated non-sheathed	2	H05V-U	1	300/500	1	0.5–1.0	+5	+40
Thermoplastic insulated and sheathed	3		1/2	300/500	2–5	1.5–35	+5	+40
Thermoplastic insulated and sheathed	4	6181Y 6192Y 6193Y	1/2	300/500	1–3	1.0–35	+5	+40
Thermoplastic insulated and sheathed	5	6242Y 6243Y	1/2	300/500	2–3	1.0–16	+5	+40
Thermoplastic insulated and sheathed	6	6192Y 6242Y	2	300/500	2	1.5–2.5	+5	+40

▼ **Table F.2** Designation system for cables complying with the European Harmonization Standard

Cable Reference

12 345 - 6 789

1	**Basic Standards**
H	Harmonized Standards
A	Authorized National Standards (derived from a harmonized cable standard)
N	Non-Authorized National Standards
2	**Rated voltage**
03	300/300 V
05	300/500 V
07	450/750 V
1	600/1000 V
3 & 4	**Insulation and sheathing material**
B	Ethylene propylene rubber (EPR)
E	Polyethylene (PE), low density (LDPE)
E2	Polyethylene, high density (HDPE)
E4	Polytetrafluoroethylene (PTFE)
E6	Ethylene tetrafluoroethylene (ETFE)
E7	Polypropylene (PP)
G	Ethylene vinyl acetate (EVA)
J	Glass fibre braid (GFB)
N	Polychloroprene (PCP)
N4	Chlorosulphonated polyethylene (CSP)
Q	Polyurethane (PU)

continues

Table F.2 *continued*

Q2	Polyethylene terephthalate (PETP)
Q4	Polyamide (PA)
R	Natural rubber
S	Silicone
T	Textile braid
V	Polyvinyl chloride (PVC)
V2	Heat-resistant polyvinyl chloride (HR PVC)
X	Cross-linked polyethylene (XLPE)
5 Special construction and shapes	
H	Flat construction with divisible cores
H2	Flat construction, non-divisible core
H5	Two or more cores twisted together, non-sheathed
6 Type of conductor	
A	Aluminium
	Copper (no code letter)
F	Flexible for movable installations (Class 5 IEC 228)
H	Highly flexible for movable installations (Class 6 IEC 228)
K	Flexible for fixed installations (Class 5 IEC 228)
R	Stranded (Class 2 IEC 228)
U	Solid (Class 1 IEC 228)
Y	Tinsel
7 Number of cores	
8 Protective conductor	
X	Without protective core
G	With protective core
9 Nominal cross-sectional area of conductors in mm²	

Additional designations	
Concentric conductors and screens	
A	Concentric aluminium conductor
C	Concentric copper conductor
A7	Aluminium/Laminate screen
C4	Overall copper braid screen
C5	Cores individually copper braid screen
C7	Lapped copper (wire, tape or strip) screen
Special components	
D3	Central strainer (textile or metallic)
D5	Central filler (not load bearing)
Armours	
Z2	Steel wire armour
Z3	Flat steel wire armour
Z4	Steel tape armour
Z5	Steel wire braid

Table F.3
Commonly used cable type reference numbers

Reference no.	Cable type
0361	Welding cable, tinned copper conductor, vulcanised rubber insulation/ tough rubber sheath. BS 638
2093Y	3-core, thermoplastic heat resisting insulation/thermoplastic heat resisting sheathed. Circular, 300/300 V BS 6500, flexible cord
218*Y	Circular, thermoplastic insulation/thermoplastic sheathed. 300/300 V BS 6500, flexible cord
2192Y	2-core, thermoplastic insulation/thermoplastic sheathed Flat 300/300 V. BS 6500, flexible cord
2213	3-core, vulcanised rubber insulation/tough rubber sheathed/cotton braid, 300/300 V BS 6500, flexible cord
318*	Circular, vulcanised rubber insulation/tough cord rubber sheathed, 300/500 V BS 6500, flexible cord
318*Y	Circular, thermoplastic insulation/thermoplastic sheathed 300/500 V BS 6500, flexible cord
318*P	Circular, vulcanised rubber insulation/PCP sheathed, 300/500 V BS 6500, flexible cord
318*TQ	Circular, EPR insulation/CSP sheathed 300/500V BS 6500, flexible cord
618*Y	Circular, thermoplastic insulation/thermoplastic sheathed 1–35 mm^2 300/500 V BS 6004 50–630 mm^2 gen in acc BS 6346, wiring cable
619*Y	Flat, thermoplastic insulation/thermoplastic sheathed 300/500 V BS 6004, wiring cable
624*Y	Flat, thermoplastic insulation – laid up in parallel with a bare CPC within one interstice/thermoplastic sheathed 300/500 V BS 6004, wiring cable
6491X	Single-core, thermoplastic insulation 1–630 mm^2. 450/750 V BS 6004
657*TQ	Circular, EPR insulation/CSP sheathed 600/1000 V BS 6883, ship-wiring cable. (circular conductors)
657*TQ(S)	As 657*TQ but with shaped conductors
658*TQ	Circular, EPR insulation/CSP bedding/galv. mild steel wire braid armour/ CSP sheathed 600/1000 V BS 6883, ship wiring cable (circular conductors)
658*TQ(S)	As 658*TQ but with shaped conductors
694*X	Circular, thermoplastic insulation/thermoplastic bedded/single steel wire armoured/thermoplastic sheathed 600/1000 V BS 6346, mains wiring cable
694*X(S)	As 694*X but with shaped conductors
H07V-R	Single-core, thermoplastic insulation, 1.5–630 mm^2, (rigid strand) 450/750 V to BS 6004
H07V-U	Single-core, thermoplastic insulation, 1.5–2.5 mm^2, (solid) 450/750 V to BS 6004

continues

▼ **Table F.3**
continued

Reference no.	Cable type
H05V-U	Single-core, thermoplastic insulation, 1.0 mm^2 (solid) 300/500 V to BS 6004
H05RR-F	Circular, vulcanised rubber insulation/tough rubber sheathed. 300/500 V to BS 6500
H03VV-F	Circular, thermoplastic insulated/thermoplastic sheathed, 300/300 V to BS 6500
H03VVH2-F	Flat 2-core, thermoplastic insulated/thermoplastic sheathed, 300/300 V to BS 6500
H05VV-F	Circular, thermoplastic insulated/thermoplastic sheathed, 300/500 V to BS 6500
H05VVH2-F	Flat, 2-core, thermoplastic insulated/thermoplastic sheathed, 300/500 V to BS 6500
H05V-K	Single-core, thermoplastic insulation 0.5–1.0 mm^2 300/500 V to BS 6500
H07V-K	Single-core, thermoplastic insulation, 1.5–2.5 mm^2, 450/750 V to BS 6004
H07RN-F	Circular, rubber insulated/rubber sheathed, 450/750 V to BS 6007

Notes:

* = number of cores (e.g. 2183Y is 3 core)

(AL) = aluminium conductors. To be added after other suffix, if used, e.g. 6491X(AL) and 6242Y(AL)

(S) = shaped conductors

(Reproduced with kind permission of Anixter (UK) Ltd.)

Table F.4 Applications of cables for fixed wiring

Type of cable	Uses	Comments
Thermoplastic, thermosetting or rubber insulated non-sheathed	In conduit, cable ducting or trunking	(i) intermediate support may be required on long vertical runs (ii) 70 °C maximum conductor temperature for normal wiring grades – including thermosetting types[4] (iii) cables run in PVC conduit shall not operate with a conductor temperature greater than 70 °C[4]
Flat thermoplastic or thermosetting insulated and sheathed	(i) general indoor use in dry or damp locations. May be embedded in plaster (ii) on exterior surface walls, boundary walls and the like (iii) overhead wiring between buildings[6] (iv) underground in conduits or pipes (v) in building voids or ducts formed in-situ	(i) additional protection may be necessary where exposed to mechanical stresses (ii) protection from direct sunlight may be necessary. Black sheath colour is better for cables in sunlight (iii) see note 4 (iv) unsuitable for embedding directly in concrete (v) may need to be hard drawn (HD) copper conductors for overhead wiring[6]
Split-concentric thermoplastic insulated and sheathed	General	(i) additional protection may be necessary where exposed to mechanical stresses (ii) protection from direct sunlight may be necessary. Black sheath colour is better for cables in sunlight
Mineral insulated	General	With overall PVC covering where exposed to the weather or risk of corrosion, or where installed underground, or in concrete ducts
Thermoplastic or XLPE insulated, armoured, thermoplastic sheathed	General	(i) additional protection may be necessary where exposed to mechanical stresses (ii) protection from direct sunlight may be necessary. Black sheath colour is better for cables in sunlight
Paper-insulated, lead sheathed and served	General, for main distribution cables	With armouring where exposed to severe mechanical stresses or where installed underground

Notes:

1 The use of cable covers (preferably conforming to BS 2484) or equivalent mechanical protection is desirable for all underground cables which might otherwise subsequently be disturbed. Route marker tape should also be installed, buried just below ground level.

2 Cables having thermoplastic insulation or sheath should preferably be installed only where the ambient temperature is above 0 °C and has been for the preceding 24 hours. Where they are to be installed during a period of low temperature, precautions should be taken to avoid risk of mechanical damage during handling. A minimum ambient temperature of 5 °C is advised in BS 7540-2:2005 for some types of thermoplastic insulated and sheathed cables.

3 Cables shall be suitable for the maximum ambient temperature and shall be protected from any excess heat produced by other equipment, including other cables.

4 Thermosetting cable types (to BS 7211 or BS 5467) can operate with a conductor temperature of 90 °C. This must be limited to 70 °C when drawn into a conduit, etc., with thermoplastic insulated conductors or connected to electrical equipment (512.1.2 and Table 52.1), or when such cables are installed in plastic conduit or trunking.
5 For cables to BS 6004, BS 6007, BS 7211, BS 6346, BS 5467 and BS 6724, further guidance may be obtained from those standards. Additional advice is given in BS 7540-2: 2005 *Guide to use of cables with a rated voltage not exceeding 450/750 V* for cables to BS 6004, BS 6007 and BS 7211.
6 Cables for overhead wiring between buildings must be able to support their self-weight and any imposed wind or ice/snow loading. A catenary support is usual but hard drawn copper types may be used.

▼ **Table F.5**
Applications of flexible cables and cords to BS 6500:2000 and BS 7919:2001

Type of flexible cord	Uses
Light thermoplastic (PVC) insulated and sheathed flexible cord	Indoors in household or commercial premises in dry situations, for light duty.
Ordinary thermoplastic (PVC) insulated and sheathed flexible cord	Indoors in household or commercial premises, including damp situations, for medium duty. For cooking and heating appliances where not in contact with hot parts. For outdoor use other than in agricultural or industrial applications. For electrically powered hand tools.
60 °C thermosetting (rubber) insulated braided twin and three-core flexible cord	Indoors in household or commercial premises where subject only to low mechanical stresses.
60 °C thermosetting (rubber) insulated and sheathed flexible cord	Indoors in household or commercial premises where subject only to low mechanical stresses. For occasional use outdoors. For electrically powered hand tools.
60 °C thermosetting (rubber) insulated oil-resisting with flame-retardant sheath	For general use, unless subject to severe mechanical stresses. For use in fixed installations where protected by conduit or other enclosures.
90 °C thermosetting (rubber) insulated HOFR sheathed	General, including hot situations, e.g. night storage heaters, immersion heaters and boilers.
90 °C heat resisting thermoplastic (PVC) insulated and sheathed	General, including hot situations, e.g. for pendant luminaires.
150 °C thermosetting (rubber) insulated and braided	For use at high ambient temperatures. For use in or on luminaires.
185 °C glass-fibre insulated single-core twisted twin and three-core	For internal wiring of luminaires only and then only where permitted by BS 4533.
185 °C glass-fibre insulated braided circular	For dry situations at high ambient temperatures and not subject to abrasion or undue flexing. For the wiring of luminaires.

Notes:
1 Cables and cords having thermoplastic insulation or sheath should preferably not be used where the ambient temperature is consistently below 0 °C. Where they are to be installed during a period of low temperature, precautions should be taken to avoid risk of mechanical damage during handling.
2 Cables and cords shall be suitable for the maximum ambient temperature and shall be protected from any excess heat produced by other equipment, including other cables.

3 For flexible cords and cables to BS 6007, BS 6141 and BS 6500 further guidance may be obtained from those standards, or from BS 7540-2:2005 *Guide to use of cables with a rated voltage not exceeding 450/750 V.*
4 When used as connections to equipment, flexible cables and cords should be of the minimum practical length to minimize danger and, in any case, of such a length that allows the protective device to operate correctly.
5 When attached to equipment, flexible cables and cords should be protected against tension, crushing, abrasion, torsion and kinking, particularly at the inlet point to the electrical equipment. At such inlet points it may be necessary to use a device which ensures that the cable is not bent to an internal radius below that given in the appropriate part of Table 4 of BS 6007. Strain relief, clamping devices or cord guards should not damage the cord.
6 Flexible cables and cords should not be used under carpets or other floor coverings, or where furniture or other equipment may rest on them. Flexible cables and cords should not be placed where there is a risk of damage from traffic passing over them.
7 Flexible cables and cords should not be used in contact with or close to heated surfaces especially if the surface approaches the upper thermal limit of the cable or cord.

F.2 British Standards

511.1
Section 522
Most cables in general use today are manufactured to a specific British Standard (or Standards) that covers the design, materials, manufacture, testing and performance of the cable. The designer of the installation has to select a cable type that is suitable for the performance required, and the relevant British Standard is usually specified.

The cable manufacturer will provide full details of cables and the standards to which they are manufactured. BS 7671 Appendix 4 indicates the relevant British Standards for each cable type in the current rating tables. A full listing of British Standards referred to in BS 7671 is provided in Appendix 1 of that Standard.

F.3 Fire stopping

527.2 All cable routes that pass between building fire zones or areas must be adequately sealed against the transmission of flames and/or smoke between zones or areas.

The specification for such fire stopping is outside the scope of this Guidance Note but should provide fire resistance to a standard at least equal to the original element of the building construction. The specification for fire stopping should be detailed by the Project Architect or Designer for all services (i.e. cables, pipes, etc.) and should comply with the requirements of the Building Regulations, as applicable. Intumescent materials are commonly used.

527.2.6 All voids within ducts, trunking, busbar trunking systems, etc., should be filled, as well as the space around such ducts, trunkings and busbar trunking systems where they pass through walls, etc. It is not usual, however, to seal the inside of conduits, except in classified hazardous areas, as required by the relevant codes or standards.

Notes on methods of support for cables, conductors and wiring systems

G

This appendix describes methods of support for cables, conductors and wiring systems which satisfy the relevant requirements of Chapter 52 of BS 7671. The use of other methods is not precluded where specified by a competent person, having due regard for protection of the cable from mechanical damage. The methods described in this appendix make no specific provisions for fire or thermal protection of the cables. All cable installations and routings must be carefully considered and protection provided as required. The advice of the local authority and the Fire Officer must be taken as necessary on the project. The Project Architect and Designer must also comply with the CDM Regulations (see Guidance Note 4 also).

G.1 Cables generally

Items **1** to **9** below are generally applicable to supports on structures which are subject only to vibration of low severity and a low risk of mechanical impact:

1 Non-sheathed cables, installed in conduit without further fixing of the cables, precautions taken against undue compression or other mechanical stressing of the insulation at the top of any vertical runs exceeding 5 m in length.

2 Cables of any type, installed in factory-made ducting or trunking without further fixing of the cables, vertical runs not exceeding 5 m in length without intermediate support.

3 Sheathed and/or armoured cables drawn into ducts formed in-situ in the building structure. The internal surfaces of the duct must be protected to prevent abrasion of the cables, especially when drawing-in. Ducts should also be adequately sealed against the spread of fire and smoke. Vertical runs not exceeding 5 m in length without intermediate support.

4 Sheathed and/or armoured cables installed in accessible positions, support by clips at spacings not exceeding the appropriate value stated in Table G.1.

5 Cables of any type, resting without fixing in horizontal runs of ducts, conduits, factory-made cable ducting or trunking.

6 Sheathed and/or armoured cables in horizontal runs which are inaccessible and unlikely to be disturbed, resting without fixing on part of a building, the surface of that part being reasonably smooth.

7 Sheathed and/or armoured cables in vertical runs which are inaccessible and unlikely to be disturbed, supported at the top of the run by a clip and a rounded support of a radius not less than the appropriate value stated in Table G.2.

8 Sheathed cables without armour in vertical runs which are inaccessible and unlikely to be disturbed, supported by the method described in **7** above; the length of run without intermediate support not exceeding 2 m for a lead sheathed cable or 5 m for a thermosetting or thermoplastic sheathed cable.

9 Thermosetting or thermoplastic sheathed cables, installation in conduit without further fixing of the cables, any vertical runs being in conduit of suitable size and not exceeding 5 m in length.

G.2 Particular applications

559.6.1.3 559.6.1.5 **10** Flexible cords used as pendants, attachment to a ceiling rose or similar accessory and lampholder by the cord grip or other method of strain relief provided in the accessories (see Table 4F3A of BS 7671 for maximum mass supportable by flexible cords).

11 Temporary installations and installations on construction sites, protection and supports so arranged that there is no appreciable mechanical strain on any cable termination or joint and the cables are protected from mechanical damage.

721.522.8.1.3 **12** Caravans, for sheathed flexible cables or cords in inaccessible spaces such as ceiling, wall and floor spaces, support at intervals not exceeding 0.25 m for horizontal runs and 0.4 m for vertical runs. In caravans, for horizontal runs of sheathed flexible cables or cords passing through floor or ceiling joists in inaccessible floor or ceiling spaces, securely bedded in thermal insulating material, no further fixing is required. (See Regulation 721.521 for details of caravan wiring systems.)

G.3 Overhead wiring

13 Cables sheathed with thermosetting or thermoplastic, supported by a separate catenary wire, either continuously bound up with the cable or attached thereto at intervals not exceeding those stated in column 2 of Table G.1 and the minimum height above ground being in accordance with Table G.3.

14 Support by a catenary wire incorporated in the cable during manufacture, the spacings between supports not exceeding those stated by the manufacturer and the minimum height above ground being in accordance with Table G.3.

15 Spans without intermediate support (e.g. between buildings) of multicore sheathed cable, with or without armour, terminal supports so arranged that no undue strain is placed upon the conductors or insulation of the cable, adequate precautions being taken against any risk of chafing of the cable sheath and the minimum height above ground and the length of such spans being in accordance with the appropriate values indicated in Table G.3. Hard drawn (HD) copper conductors may be necessary for longer spans. Cable manufacturers' advice should be complied with.

16 Bare or insulated conductors of an overhead line for distribution between a building and a remote point of utilization (e.g. another building) supported on insulators, the lengths of span and heights above ground having the appropriate values indicated in Table G.3 or otherwise installed in accordance with the Electricity Safety, Quality and Continuity Regulations 2002 (as amended).

17 Spans without intermediate support (e.g. between buildings) and which are in situations inaccessible to vehicular traffic, insulated or multicore sheathed cables installed in heavy gauge galvanized steel conduit, the length of span and height above ground being in accordance with Table G.3, provided that the conduit shall be earthed in accordance with Parts 4 and 5 of BS 7671, must be securely fixed at the ends of the span and not be jointed in its span.

Table G.1 Spacings of supports for cables in accessible positions where the entire support is derived from the clips

Overall diameter of cable, d* (mm)	Maximum spacings of clips (mm)							
	Non-armoured thermosetting, plastics or lead sheathed cables				Armoured cables		Mineral insulated copper sheathed cables	
	Generally		In caravans					
	Horizontal†	Vertical†	Horizontal†	Vertical†	Horizontal†	Vertical†	Horizontal†	Vertical†
1	2	3	4	5	6	7	8	9
d ≤ 9	250	400	250	400	–	–	600	800
9 < d ≤ 15	300	400	as above		350	450	900	1200
15 < d ≤ 20	350	450	as above		400	550	1500	2000
20 < d ≤ 40	400	550	as above		450	600	2000	3000

Notes:

- For the spacing of supports for cables having an overall diameter exceeding 40 mm, or conductors of cross-sectional area 300 mm^2 and larger, the manufacturer's recommendations should be observed.
- The spacings given in this table are maxima and for good workmanship in certain circumstances they may need to be reduced.

* For flat cables taken as the dimension of the major axis.

† The spacings stated for horizontal runs may be applied also to runs at an angle of more than 30° from the vertical. For runs at an angle of 30° or less from the vertical, the vertical spacings are applicable.

▼ Table G.2
Minimum internal radii of bends in cables for fixed wiring

Insulation	Finish	Overall diameter*	Factor to be applied to overall diameter of cable to determine minimum internal radius of bend
Thermosetting or thermoplastic (circular, or circular stranded copper or aluminium conductors)	Non-armoured	Not exceeding 10 mm	3(2)†
		Exceeding 10 mm but not exceeding 25 mm	4(3)†
		Exceeding 25 mm	6
	Armoured	Any	6
Thermosetting or thermoplastic (solid aluminium or shaped copper conductors)	Armoured or non-armoured	Any	8
Mineral	Copper sheath with or without covering	Any	6‡
Flexible cables	Sheathed	Any	No specific provision but no tighter than the equivalent sized non-armoured cable**

Notes:

* For flat cables the diameter refers to the major axis.

† The figure in brackets relates to single-core circular conductors of stranded construction installed in conduit, ducting or trunking.

‡ For mineral insulated cables, the bending radius should normally be limited to a minimum of 6 times the diameter of the bare copper sheath, as this will allow further straightening and reworking if necessary. However, cables may be bent to a radius not less than 3 times the cable diameter over the copper sheath, provided that the bend is not reworked.

** Flexible cables can be damaged by too tight, or repeated bending.

▼ **Table G.3** Maximum lengths of span and minimum heights above ground for overhead wiring between buildings, etc.

Type of system	Maximum length of span (m)	Minimum height of span above ground (m)†		
		At road crossings	In positions accessible to vehicular traffic, other than crossings	In positions inaccessible to vehicular traffic*
1	2	3	4	5
Cables sheathed with thermoplastic or having an oil-resisting and flame-retardant (HOFR) sheath without intermediate support.	3	5.8	5.8	3.5
Cables sheathed with thermoplastic or having an oil-resisting and flame-retardant (HOFR) sheath, in heavy gauge steel conduit of a diameter not less than 20 mm and not jointed in its span.	3	5.8	5.8	3
Bare or covered overhead lines supported by insulators without intermediate support.	30	5.8	5.8	5.2
Cables sheathed with thermoplastic or having an oil-resisting and flame-retardant (HOFR) sheath, supported by a catenary wire.	No limit	5.8	5.8	3.5
Overhead cable incorporating a catenary wire.	Subject to Item 14	5.8	5.8	3.5

Notes:

* This column is not applicable in agricultural premises.

† In some special cases, such as in yacht marinas or where large cranes are present, it will be necessary to increase the height of span above ground over the minimum given in the table. It is preferable to use underground cables in such locations.

▶ Schedule 2 of the Electricity Safety, Quality and Continuity Regulations 2002 should be consulted for the minimum height above ground of overhead lines.

G.4 Conduit and cable trunking

18 Rigid conduit supported in accordance with Table G.4.
19 Cable trunking supported in accordance with Table G.5.
20 Conduit embedded in the material of the building, suitably treated against corrosion if necessary.
21 Pliable conduit embedded in the material of the building or in the ground, or supported in accordance with Table G.4.

▼ **Table G.4**
Spacings of supports for conduits

Nominal diameter of conduit, d (mm)	Maximum distance between supports (m)					
	Rigid metal		Rigid insulating		Pliable	
	Horizontal	Vertical	Horizontal	Vertical	Horizontal	Vertical
$d \leq 16$	0.75	1.0	0.75	1.0	0.3	0.5
$16 < d \leq 25$	1.75	2.0	1.5	1.75	0.4	0.6
$25 < d \leq 40$	2.0	2.25	1.75	2.0	0.6	0.8
$d > 40$	2.25	2.5	2.0	2.0	0.8	1.0

Notes:

1 The figures given above are the maximum spacings.
2 Conduit boxes supporting luminaires or electrical accessories will require separate fixing.
3 Plastic conduits will require closer spacing of fixings in areas of high ambient temperature and note should be taken of the manufacturer's declared maximum temperature.
4 The spacings in the table allow for maximum fill of cables in compliance with Appendix A and thermal limits to the relevant British Standards. They assume that the conduit is not exposed to other mechanical stress.
5 A flexible conduit should be of such length that it does not need to be supported in its run.

▼ **Table G.5**
Spacings of supports for cable trunking

Cross-sectional area of trunking, A (mm²)	Maximum distance between supports (m)			
	Metal		Insulating	
	Horizontal	Vertical	Horizontal	Vertical
$300 < A \leq 700$	0.75	1.0	0.5	0.5
$700 < A \leq 1500$	1.25	1.5	0.5	0.5
$1500 < A \leq 2500$	1.75	2.0	1.25	1.25
$2500 < A \leq 5000$	3.0	3.0	1.5	2.0
$A > 5000$	3.0	3.0	1.75	2.0

Notes:

1 The spacings in the table allow for maximum fill of cables in compliance with Appendix A and thermal limits to the relevant British Standards. They assume that the trunking is not exposed to other mechanical stress.
2 The above figures do not apply to lighting suspension trunking, or where specially strengthened couplers are used.
3 Plastic trunking will require closer spacing of fixings in areas of high ambient temperature and note should be taken of the manufacturer's declared maximum temperature.

G.5 Conduit bends

The radius of every conduit bend must be such as to allow compliance with the minimum bending radii of cables installed in the conduit (e.g. see Table G.2). In addition, the inner radius of the bend should not be less than 2.5 times the outside diameter of the conduit.

G.6 Spacing of supports for busbar trunking systems

There is no standard arrangement for the spacing of supports for busbar trunking systems and therefore the manufacturer will provide details of the method of mounting. Supports, where applicable, must be located in specific positions using the spacing dimensions given in the installation instructions.

Different attitudes or mounting arrangements can change the orientation of the conductors within the busbar trunking system and may affect the current-carrying capacity of the system. Where derating is necessary, the manufacturer will provide details of the corresponding mounting factor (k_β) to determine the allowable current of the system using the following formula: Appx 8

$$I_z = k_\beta \times I_n$$

where:

I_z is the allowable current
k_β is the mounting factor
I_n is the declared rated current.

Maximum demand and diversity

H

This appendix gives some information on the determination of the maximum demand for an installation and includes the current demand to be assumed for commonly used equipment. It also includes some notes on the application of allowances for diversity. 311.1

The information and values given in this appendix are intended only for guidance because it is impossible to specify the appropriate allowances for diversity for every type of installation and such allowances call for special knowledge and experience. The figures given in Table H.2, therefore, may be increased or decreased as decided by the competent person responsible for the design of the installation concerned. For blocks of residential dwellings, large hotels, industrial and large commercial and office premises, the allowances should be assessed by a competent person.

The current demand of a final circuit is determined by adding the current demands of all points of utilization and equipment in the circuit and, where appropriate, making an allowance for diversity. Typical current demands to be used for this summation are given in Table H.1. For the standard circuits using BS 1363 socket-outlets detailed in Appendix C, an allowance for diversity has been taken into account and no further diversity should be applied.

▼ **Table H.1**
Current demand to be assumed for points of utilization and current-using equipment

Point of utilization or current-using equipment	Current demand to be assumed
Socket-outlets other than 2 A socket-outlets and 13 A socket-outlets (note 1)	Rated current
2 A socket-outlets	At least 0.5 A
Lighting outlet (note 2)	Current equivalent to the connected load, with a minimum of 100 W per lampholder
Electric clock, electric shaver supply unit (complying with BS EN 61558-2-5), shaver socket-outlet (complying with BS 4573), bell transformer, and current-using equipment of a rating not greater than 5 VA	May be neglected
Household cooking appliance	The first 10 A of the rated current plus 30% of the remainder of the rated current plus 5 A if a socket-outlet is incorporated in the control unit
All other stationary equipment	British Standard rated current, or normal current

Notes:

1 See Appendix C for the design of standard circuits using socket-outlets to BS 1363-2 and BS EN 60309-2 (BS 4343).

2 Final circuits for discharge lighting shall be arranged so as to be capable of carrying the total steady current, viz. that of the lamp(s) and any associated controlgear and also their harmonic currents. Where more exact information is not available, the demand in volt-amperes is taken as the rated lamp watts multiplied by not less than 1.8. This multiplier is based upon the assumption that the circuit is corrected to a power factor of not less than 0.85 lagging, and takes into account controlgear losses and harmonic current.

The current demand of a distribution system or distribution circuit supplying a number of final circuits may be assessed by using the allowances for diversity given in Table H.2, which are applied to the total current demand of all the equipment supplied by that circuit and not by adding the current demands of the individual final circuits obtained as outlined above. In Table H.2 the allowances are expressed either as percentages of the current demand or, where followed by the letters f.l., as percentages of the rated full load current of the current-using equipment. The current demand for any final circuit which is a standard circuit arrangement complying with Appendix C is the rated current of the overcurrent protective device of that circuit.

An alternative method of assessing the current demand of a circuit supplying a number of final circuits is to add the diversified current demands of the individual circuits then apply a further allowance for diversity but with this method the allowances given in Table H.2 should not be used, the values to be chosen being the responsibility of the installation designer.

The use of other methods of determining maximum demand is not precluded where specified by a competent person.

After the design currents for all the circuits have been determined, enabling the conductor sizes to be chosen, it is necessary to check that the design complies with the requirements of Part 4 of BS 7671 and that the limitation on voltage drop is met.

▼ **Table H.2** Allowances for diversity

Purpose of final circuit fed from conductors or switchgear to which diversity applies	Type of premises: Individual household installations including individual dwellings of a block	Small shops, stores, offices and business premises	Small hotels, boarding houses, guest houses, etc.
1 Lighting	66% of total current demand	90% of total current demand	75% of total current demand
2 Heating and power (but see 3 to 8 below)	100% of total current demand up to 10 A + 50% of any current demand in excess of 10 A	100% f.l. of largest appliance + 75% f.l. of remaining appliances	100% f.l. of largest appliance + 80% f.l. of 2nd largest appliance + 60% f.l. of remaining appliances
3 Cooking appliances	10 A + 30% f.l. of connected cooking appliances in excess of 10 A + 5 A if socket-outlet incorporated in control unit	100% f.l. of largest appliance + 80% f.l. of 2nd largest appliance + 60% f.l. of remaining appliances	100% f.l. of largest appliance + 80% f.l. of 2nd largest appliance + 60% f.l. of remaining appliances
4 Motors (other than lift motors, which are subject to special consideration)	Not applicable	100% f.l. of largest motor + 80% f.l. of 2nd largest motor + 60% f.l. of remaining motors	100% f.l. of largest motor + 50% f.l. of remaining motors
5 Water-heaters (instantaneous type)*	100% f.l. of largest appliance + 100% f.l. of 2nd largest appliance + 25% f.l. of remaining appliances	100% f.l. of largest appliance + 100% f.l. of 2nd largest appliance + 25% f.l. of remaining appliances	100% f.l. of largest appliance + 100% f.l. of 2nd largest appliance + 25% f.l. of remaining appliances
6 Water-heaters (thermostatically controlled)		No diversity allowable†	
7 Floor warming installations		No diversity allowable†	
8 Thermal storage space heating installations		No diversity allowable†	
9 Standard arrangement of final circuits in accordance with Appendix C	100% of current demand of largest circuit + 40% of current demand of every other circuit	100% of current demand of largest circuit + 50% of current demand of every other circuit	
10 Socket-outlets other than those included in 9 above and stationary equipment other than those listed above	100% of current demand of largest point of utilization + 40% of current demand of every other point of utilization	100% of current demand of largest point of utilization + 75% of current demand of every other point of utilization	100% of current demand of largest point of utilization + 75% of current demand of every other point in main rooms (dining rooms, etc.) + 40% of current demand of every other point of utilization

Notes:

* In this context an instantaneous water-heater is deemed to be a water-heater of any loading which heats water only while the tap is turned on and therefore uses electricity intermittently.

† It is important to ensure that the distribution boards, etc., are of sufficient rating to take the total load connected to them without the application of any diversity.

Minimum separating distances between electricity supply cables and telecommunications or control cables

I

The adverse effects of electromagnetic interference (EMI) on an installation and its equipment are a problem that many have experienced. Designers and installers faced with this phenomenon have, in the past, overcome the problems in a number of ways, e.g. use of filtering devices, earthing and bonding, and segregation or separation of cables.

However, to comply with a European Directive on EMC, manufacturers and installers must ensure that equipment, when installed, will:

- not interfere with other equipment in the vicinity (e.g. it shall limit its emissions) and 515.3.2
- be able to withstand likely interference levels present in the vicinity (e.g. it shall have sufficient immunity). 515.3.1

Regulation 515.3 of BS 7671 outlines the requirements for emission of electromagnetic influences and interference from such influences but does not give any guidance on how these should be achieved.

To comply with the statutory requirements of the Electromagnetic Compatibility Regulations 2005, only equipment manufactured so as to comply with the relevant standards for protection against interference and, where available, for immunity must be used, e.g.:

- BS EN 55011:2007 for Industrial, scientific and medical equipment
- BS EN 55015:2006 for Electrical lighting and similar
- BS EN 55022:2006 (and BS EN 55024:1998 +A2:2003) for Information technology equipment.

Where equipment does not have a specific standard, or where a specific standard covers only emissions and not immunity, equipment should conform to the appropriate part of BS EN 61000 series.

Further information is also available in *British Standards on segregation of cables*, e.g. BS 6701:2004 for the installation of telecommunication cables.

The avoidance of problems should be given due regard at the design stage and segregation is considered to be a way of overcoming the problems of EMI.

The segregation details given in Table I.1 apply only to cases where adjacent runs of both power and signal cables are envisaged.

With regard to the values in the table, it is important to note that these are minimum distances and it is assumed the equipment connected to the IT cables meets the generic Standards, which may well not be the case for existing equipment. Moreover, the manufacturers of electronic equipment may, in addition to any requirements for cable separation on the grounds of safety, specify that greater separations are used for installations including their equipment.

A major factor affecting the performance of the cables in EMC terms is the quality of termination of the screen. It is therefore important that the correct termination is used.

The values apply to up to 100 m parallel run and with up to 125 A flowing in the power cable.

▼ **Table I.1**
Minimum segregation distances for power and information technology cables that run parallel

	Minimum segregation distance (mm)		
	Without divider or non-metallic divider[1]	Aluminium divider	Steel divider
Unscreened power cable and IT cable	200	100	50
Unscreened power cable and screened IT cable	50	20	5
Screened power cable and unscreened IT cable	30	10	2
Screened power cable and screened IT cable[2]	0	0	0

Notes:

1 It is assumed that in case of a metallic divider, the design of the cable management system will achieve a screening attenuation related to the material used for the divider.

2 The screened IT cables shall comply with BS EN 50288 series.

For horizontal cabling the following applies:

- If the full length is no greater than 35 m no separation is required.
- For lengths greater than 35 m the separation distances apply to the full length excluding the final 15 m attached to the outlet.

The second amendment to BS 7671:1992 deleted circuit categories and substituted 'Voltage Bands'. Band I effectively covers ELV circuits (including SELV) and Band II effectively covers LV circuits. These bands only relate to separation for electrical safety and do not relate to separation for electromagnetic interference requirements. It is still possible in some cases to have a cable for a 230 V power circuit near to a cable for an intruder alarm. This method of installation would be unlikely to meet the requirements of the legislation and interference on the security system could be experienced.

The need for system cables to be separated is in conflict with the usual building design for the minimum space to be allowed for cable routes but generally separation between system cables should be as large as possible, though available space will always be a limiting factor. Especially sensitive systems or systems that will emit interference should be identified and adequate provisions made in the design with the advice of the system manufacturer. It may be that building designers will have to allow more space for electrical cable routes and risers than they would wish. All cable systems should be identified as necessary for ease of future modification and maintenance.

High voltage system cables should be segregated from cables of other systems and clearly identified as a general safety precaution, as well as for EMC reasons. If separation cannot be achieved, some form of protective screen or barrier may be required. Table I.2 recommends minimum separation distances between systems.

▼ **Table I.2**
Separation of systems

Voltage to earth	Normal separation distance	Exceptions to normal separation distance, plus conditions to exception
Exceeding 600 V a.c. and 900 V d.c.	150 mm	50 mm, as long as a divider that maintains separation of 50 mm is inserted between two sets of cables. The divider should be made of a rigid, non-conducting, non-flammable material

Permitted protective conductor currents J

Protective conductor/touch current measurement is an alternative to the in-service insulation test for use if the insulation resistance test either cannot be carried out or gives suspect test results. The current is measured from live parts to earth for Class I equipment, or from live parts to accessible surfaces of Class II equipment.

The current is to be measured within 5 s after the application of the test voltage and must not exceed the values in Table J.1.

For practical purposes the test voltage is the supply voltage.

▼ Table J.1
Measured protective conductor/touch current

Appliance class	Maximum current[1]
Portable or hand-held Class I equipment	0.75 mA
Class I heating appliances	0.75 mA or 0.75 mA per kW, whichever is the greater, with a maximum of 5 mA
Other Class I equipment	3.5 mA
Class II equipment	0.25 mA
Class III equipment	0.5 mA

Notes:

1 The values specified above are doubled:

- if the appliance has no control device other than a thermal cut-out, a thermostat without an 'off' position or an energy regulator without an 'off' position.
- if all control devices have an 'off' position with a contact opening of at least 3 mm and disconnection in each pole.

2 Equipment with a protective conductor current designed to exceed 3.5 mA shall comply with the requirements of Section 543.7 of BS 7671.

3 The nominal test voltage is:

- 1.06 times rated voltage, or 1.06 times the upper limit of the rated voltage range, for appliances for d.c. only, for single-phase appliances and for three-phase appliances which are also suitable for single-phase supply, if the rated voltage or the upper limit of the rated voltage range does not exceed 250 V;
- 1.06 times rated line voltage divided by 1.73, or 1.06 times the upper limit of the rated voltage range divided by 1.73 for other three-phase appliances.

Standard symbols, units and graphical symbols for general electrical purposes

K

The following symbols are extracted mainly from IEC 60617, supplemented by references from other Standards.

K.1 General symbols

V	volts
A	amperes
Hz	hertz
W	watts
kW	kilowatts
F	farads
p.u.	per unit
ph	phase
p.f.	power factor
L	line
N	neutral
h	hours
min	minutes
s	seconds

⎓ direct current (d.c.)

~ alternating current (a.c.)

2 ~ two-phase alternating current

2N ~ two-phase alternating current with neutral

3 ~ three-phase alternating current

3N ~ three-phase alternating current with neutral

IPXX IP number (see Appendix B)

 fault (indication of assumed fault location)

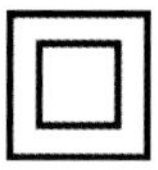 Class II appliance (Equipment in which protection against electric shock does not rely on basic insulation only but in which additional safety precautions such as supplementary insulation are provided, there being no provision for the connection of exposed metalwork of the equipment to a protective conductor and no reliance upon precautions to be taken in the fixed wiring of the installation (see BS 2754)).

 Class III appliance (Equipment in which protection against electric shock relies on supply at SELV and in which voltages higher than those of SELV are not generated (see BS 2754)).

 Safety isolating transformer. Class III equipment must be supplied from a safety isolating transformer to BS EN 61558-2-6. The safety isolating transformer will have this identifying mark upon it.

 Isolating transformer.

 Protective earth, general symbol (preferred to ).

In the International System of units (known as SI), there are seven base units, as shown below; other quantities are derived from these (see K.1.1).

Quantity	Name of base unit	Unit symbol
Length	metre	m
Mass	kilogram	kg
Time	second	s
Electric current	ampere	A
Thermodynamic temperature	kelvin	K
Amount of substance	mole	mol
Luminous intensity	candela	cd

Multiples and sub-multiples of quantities*								
10^{18}	exa	E				10^{-3}	milli	m
10^{15}	peta	P	10^{2}	hecto	h	10^{-6}	micro	μ
10^{12}	tera	T	10^{1}	deca	da	10^{-9}	nano	n
10^{9}	giga	G	10^{-1}	deci	d	10^{-12}	pico	p
10^{6}	mega	M	10^{-2}	centi	c	10^{-15}	femto	f
10^{3}	kilo	k				10^{-18}	atto	a

* Powers in steps of 3 are preferred but some others have common usage (e.g. centimetre, cm; decibel, dB).

K.1.1 SI derived units

The units of all physical quantities are derived from the base and supplementary SI units and certain of them have been named. These, together with some common compound units, are given here:

Quantity	Unit name	Unit symbol	SI units
Force	newton	N	kg m/s^2
Energy	joule	J	N m
Power	watt	W	J/s
Pressure, stress	pascal	Pa	N/m^2
Electric potential	volt	V	J/C, W/A
Electric charge, electric flux	coulomb	C	A s
Magnetic flux	weber	Wb	V s
Magnetic flux density	tesla	T	Wb/m^2
Resistance	ohm	Ω	V/A
Conductance	siemens	S	A/V
Capacitance	farad	F	C/V
Inductance	henry	H	Wb/A
Celsius temperature	degree Celsius	°C	K
Frequency	hertz	Hz	s^{-1}
Luminous flux	lumen	lm	cd sr
Illuminance	lux	lx	lm/m^2
Activity (radiation)	becquerel	Bq	s^{-1}
Absorbed dose	gray	Gy	J/kg
Dose equivalent	sievert	Sv	J/kg
Mass density	kilogram per cubic metre	kg/m^3	
Torque	newton metre	N m	
Electric field strength	volt per metre	V/m	
Magnetic field strength	ampere per metre	A/m	
Thermal conductivity	watt per metre kelvin	W m^{-1}K^{-1}	
Luminance	candela per square metre	cd/m^2	

K.2 Symbols for use in schematic wiring diagrams

Transformer, general symbol

General wiring, with joint

Identification of specific conductors

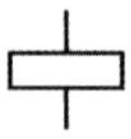
Operating device, general symbol (coil)

Make contact, normally open

Break contact, normally closed

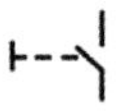
Manually operated switch, general symbol

Three-phase winding, delta

Three-phase winding, star

Changer, general symbol
Convertor, general symbol

Notes:

1 If the direction of change is not obvious, it may be indicated by an arrowhead on the outline of the symbol

2 A symbol or legend indicating the input or output quantity, waveform, etc., may be inserted in each half of the general symbol to show the nature of the change.

Rectifier

Inverter

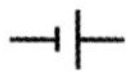
Battery of primary or secondary cells

Fuse link, rated current in amperes

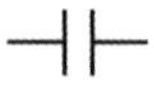
Capacitor, general symbol

Inductor coil, winding or choke

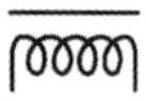
Inductor coil, winding or choke with magnetic core

Semiconductor diode, general symbol

K.2.1 Making and breaking current

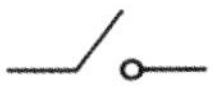 Switch

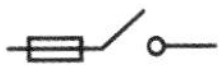 Switch–fuse

 Fuse–switch

K.2.2 Isolating

 Isolator (disconnector), general symbol

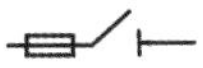 Disconnector–fuse (fuse combination unit)

 Fuse–disconnector

 Circuit-breaker suitable for isolation

K.2.3 Making, breaking and isolating

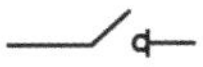 Switch–disconnector

 Switch–disconnector–fuse (fuse combination unit)

 Fuse–switch–disconnector

K.2.4 Meters

Voltmeter

Ammeter

Integrating instrument or energy meter
* Function
Wh = watt-hour
VArh = volt ampere reactive hour

K.3 Location symbols for installations

Machine, general symbol
* Function
M = Motor
G = Generator

 Motor starter, general symbol

 Star-delta starter

 Socket-outlet, general symbol

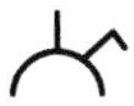 Switched socket-outlet

	Switch, general symbol
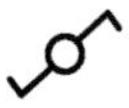	2-way switch, single-pole
	Intermediate switch
	Pull switch, single-pole
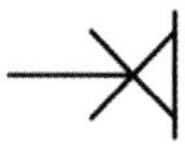	Lighting outlet position, general symbol
	Fluorescent luminaire, general symbol
	Emergency lighting luminaire (or special circuit)
	Self-contained emergency lighting luminaire
	Push button with indicator lamp
	Lamp or signal lamp, general symbol
	Clock, general symbol
	Acoustic signalling device, general symbol (e.g. bell)
	Buzzer
	Telephone handset, general symbol
	Microphone
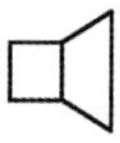	Loudspeaker
	Antenna, general symbol
	Static generator

Addresses of associated bodies and industry contacts

L

Amicus
Hayes Court
West Common Road
Hayes
Kent BR2 7AU

Tel: 0845 850 4242
Web: www.amicustheunion.org

Association of Consulting Engineers (ACE)
Alliance House
12 Caxton Street
London SW1H 0QL

Tel: 020 7222 6557
Fax: 020 7222 0750
Web: www.acenet.co.uk

Association of Manufacturers of Domestic Appliances (AMDEA)
Rapier House
40–46 Lamb's Conduit Street
London WC1N 3NW

Tel: 020 7405 0666
Fax: 020 7405 6609
Web: www.amdea.org.uk

ASTA Certification Services (also BEAB)
(now part of Intertek)
ASTA House
Chestnut Field
Rugby
Warwickshire CV21 2TL

Tel: 01788 578435
Fax: 01788 573605
Web: www.astabeab.co.uk

British Approvals Service for Cables (BASEC)
23 Presley Way
Crownhill, Milton Keynes
Buckinghamshire MK8 0ES

Tel: 01908 267300
Fax: 01908 267255
Web: www.basec.org.uk

British Cables Association (BCA)
Web: www.bcauk.org

British Electrotechnical & Allied Manufacturers' Association (BEAMA)
Westminster Tower
3 Albert Embankment
London SE1 7SL

Tel: 020 7793 3000
Fax: 020 7793 3003
Web: www.beama.org.uk

British Electrotechnical Approvals Board (BEAB)
(see ASTA Certification Services above)

British Standards Institution (BSI)
389 Chiswick High Road
London W4 4AL

Tel: 020 8996 9001
Fax: 020 8996 7401
Web: www.bsi-global.com

Building Services Research and Information Association (BSRIA)
Old Bracknell Lane West
Bracknell
Berkshire RG12 7AH

Tel: 01344 465600
Fax: 01344 465626
Web: www.bsria.co.uk

Chartered Institution of Building Services Engineers (CIBSE)
Delta House
222 Balham High Road
London SW12 9BS

Tel: 020 8675 5211
Fax: 020 8675 5449
Web: www.cibse.org

City and Guilds (C&G)
1 Giltspur Street
London EC1A 9DD

Tel: 020 7294 2800
Fax: 020 7294 2400
Web: www.city-and-guilds.co.uk

Copper Development Association
5 Grovelands Business Centre
Boundary Way
Hemel Hempstead
Hertfordshire HP2 7TE

Fax: 01442 275716
Web: www.cda.org.uk

Department for Business, Enterprise and Regulatory Reform
(formerly DTI)
Ministerial Correspondence Unit
Department for Business, Enterprise and Regulatory Reform
1 Victoria Street
London SW1H 0ET

Tel: 020 7215 5000
Fax: 020 7215 0105
Web: www.berr.gov.uk

Electrical Contractors' Association (ECA)
ESCA House
34 Palace Court
London W2 4HY

Tel: 020 7313 4800
Fax: 020 7221 7344
Web: www.eca.co.uk

Electrical Safety Council
18 Buckingham Gate
London SW1E 6LB

Tel: 0870 040 0561
Fax: 0870 040 0560
Web: www.electricalsafetycouncil.org.uk

Energy Industries Council (EIC)
Newcombe House
45 Notting Hill Gate
London W11 3LQ

Tel: 020 7221 2043
Fax: 020 7221 8813
Web: www.the-eic.com

Energy Networks Association
18 Stanhope Place
Marble Arch
London W2 2HH

Tel: 020 7706 5100
Web: www.energynetworks.org

Engineering Equipment & Material Users' Association
10–12 Lovat Lane
London EC3R 8DN

Tel: 020 7621 0011
Fax: 020 7621 0022
Web: www.eemua.org

ERA Technology Ltd
Cleeve Road
Leatherhead
Surrey KT22 7SA

Tel: 01372 367000
Fax: 01372 367099
Web: www.era.co.uk

Federation of the Electronics Industry
Russell Square House
10–12 Russell Square
London WC1B 5EE

Tel: 020 7331 2000
Fax: 020 7331 2040
Web: www.fei.org.uk

Fibreoptic Industry Association
The Manor House
Buntingford
Hertfordshire SG9 9AB

Tel: 01763 273039
Fax: 01763 273255
Web: www.fia-online.co.uk

GAMBICA Association Ltd
(Association for the Instrumentation, Control and Automation Industry in the UK)
Broadwall House
21 Broadwall Street
London SE1 9PL

Tel: 020 7642 8080
Fax: 020 7642 8096
Web: www.gambica.org.uk

Health and Safety Executive (HSE)
Tel: 0845 345 0055
Web: www.hse.gov.uk

HSE Books
PO Box 1999
Sudbury CO10 2WA

Tel: 01787 881165
Web: www.hsebooks.com

Institution of Engineering and Technology (IET)
Michael Faraday House
Stevenage
Hertfordshire SG1 2AY

Tel: 01438 313 311
Fax: 01438 765 526
Web: www.theiet.org

Joint Industry Board for the Electrical Contracting Industry
Kingswood House
47/51 Sidcup Hill
Sidcup
Kent DA14 6HP

Tel: 020 8302 0031
Fax: 020 8309 1103
Web: www.jib.org.uk

Lighting Association Ltd
Stafford Park 7
Telford
Shropshire TF3 3BQ

Tel: 01952 290905
Fax: 01952 290906
Web: www.lightingassociation.com

Lighting Industry Federation Ltd (LIF)
Ground Floor
Westminster Tower
3 Albert Embankment
London SE1 7SL

Tel: 020 7793 3020
Fax: 020 7793 3003
Web: www.lif.co.uk

London District Surveyors Association
Web: www.londonbuildingcontrol.org.uk

National Joint Utilities Group (NJUG)
28 Broadway
Westminster
London SW1H 9JX

Tel: 020 7340 1423
Web: www.njug.org.uk

NICEIC Group Ltd
Warwick House
Houghton Hall Park
Houghton Regis, Dunstable
Bedfordshire LU5 5ZX

Tel: 0870 013 0382
Fax: 01582 53090
Web: www.niceic.org.uk

Royal Institute of British Architects (RIBA)
66 Portland Place
London W1B 1AD

Tel: 020 7307 3700
Fax: 020 7436 9112
Web: www.ribafind.org

Safety Assessment Federation
Unit 4, First Floor
70 South Lambeth Road
Vauxhall
London SW8 1RL

Tel: 020 7582 3208
Fax: 020 7735 0286
Web: www.safed.co.uk

SELECT
The Walled Garden
Bush Estate
Midlothian
Scotland EH26 0SB

Tel: 0131 445 5577
Fax: 0131 445 5548
Web: www.select.org.uk

United Kingdom Accreditation Service (UKAS)
21–47 High Street
Feltham
Middlesex TW13 4UN

Tel: 020 8917 8400
Fax: 020 8917 8500
Web: www.ukas.com

Index

E

M

N

O

P

The IEE

IEE Wiring Regulations and associated publications

The IEE prepares regulations for the safety of electrical installations for buildings, the *IEE Wiring Regulations* (BS 7671 *Requirements for Electrical Installations*), which have now become the standard for the UK and many other countries. It also recommends, internationally, the requirements for ships and offshore installations. The IEE provides guidance on the application of the installation regulations through publications focused on the various activities from design of the installation through to final test and then maintenance. This includes a series of eight Guidance Notes, two Codes of Practice and Model Forms for use in Wiring Installations.

Requirements for Electrical Installations BS 7671:2008 (IEE Wiring Regulations, 17th Edition)
Order book PWR1700B Paperback 2008
ISBN: 978-0-86341-844-0 **£65**

On-Site Guide (BS 7671:2008 17th Edition)
Order book PWGO170B 188pp Paperback 2008
ISBN: 978-0-86341-854-9 **£20**

Wiring Matters Magazine **FREE**
If you wish to receive a FREE copy or advertise in Wiring Matters please visit www.theiet.org/wm

IEE Guidance Notes

A series of Guidance Notes has been issued, each of which enlarges upon and amplifies the particular requirements of a part of the IEE Wiring Regulations.

Guidance Note 1: Selection & Erection of Equipment, 5th Edition
Order book PWG1170B 216pp Paperback 2009
ISBN: 978-0-86341-855-6 **£30**

Guidance Note 2: Isolation & Switching, 5th Edition
Order book PWG2170B 74pp Paperback 2009
ISBN: 978-0-86341-856-3 **£25**

Guidance Note 3: Inspection & Testing, 5th Edition
Order book PWG3170B 128pp Paperback 2008
ISBN: 978-0-86341-857-0 **£25**

Guidance Note 4: Protection Against Fire, 5th Edition
Order book PWG4170B 98pp Paperback 2009
ISBN: 978-0-86341-858-7 **£25**

Guidance Note 5: Protection Against Electric Shock, 5th Edition
Order book PWG5170B 115pp Paperback 2009
ISBN: 978-0-86341-859-4 **£25**

Guidance Note 6: Protection Against Overcurrent, 5th Edition
Order book PWG6170B 113pp Paperback 2009
ISBN: 978-0-86341-860-0 **£25**

Guidance Note 7: Special Locations, 3rd Edition
Order book PWG7170B 142pp Paperback 2009
ISBN: 978-0-86341-861-7 **£25**

Guidance Note 8: Earthing & Bonding, 1st Edition
Order book PWRG0241 168pp Paperback 2007
ISBN: 978-0-86341-616-3 **£25**

Continued overleaf ▶

IET
Collective **inspiration**

Order Form

How to order

BY PHONE:
+44 (0)1438 767328
BY FAX:
+44 (0)1438 767375
BY EMAIL:
sales@theiet.org
BY POST:
The Institution of Engineering and Technology, PO Box 96, Stevenage SG1 2SD, UK
OVER THE WEB:
www.theiet.org/books

*Postage/Handling: Postage within the UK is £3.50 for any number of titles. Outside UK (Europe) add £5.00 for first title and £2.00 for each additional book. Rest of World add £7.50 for the first book and £2.00 for each additional book. Books will be sent via air-mail. Courier rates are available on request, please call +44 (0) 1438 767328 or email sales@theiet.org for rates.

** To qualify for discounts, member orders must be placed directly with the IET.

GUARANTEED RIGHT OF RETURN:
If at all unsatisfied, you may return book(s) in new condition within 30 days for a full refund. Please include a copy of the invoice.

DATA PROTECTION:
The information that you provide to the IET will be used to ensure we provide you with products and services that best meet your needs. This may include the promotion of specific IET products and services by post and/or electronic means. By providing us with your email address and/or mobile telephone number you agree that we may contact you by electronic means. You can change this preference at any time by visiting www.theiet.org/my.

Details

Name:

Job Title:

Company/Institution:

Address:

Postcode: Country:

Tel: Fax:

Email:

Membership No (if Institution member):

Payment methods

☐ By **cheque** made payable to The Institution of Engineering and Technology

☐ By **credit/debit card:**

☐ Visa ☐ Mastercard ☐ American Express ☐ Maestro Issue No:____________________

Valid from: ☐☐ ☐☐ Expiry Date: ☐☐ ☐☐ Card Security Code: ☐☐☐☐
(3 or 4 digits on reverse of card)

Card No: ☐☐☐☐ ☐☐☐☐ ☐☐☐☐ ☐☐☐☐

Signature____ Date ______________________
(Orders not valid unless signed)

Cardholder Name:

Cardholder Address:

Town: Postcode:

Country:

☐ By official **company purchase order** (please attach copy)

EU VAT number:____

Ordering information

Quantity	Book No.	Title/Author	Price (£)
		Subtotal	
		- Member discount**	
		+ Postage /Handling*	
		+ VAT (if applicable)	
		Total	

Notes

Notes

Notes

Notes